Apollo 11

The NASA Mission Reports

Compiled from the NASA archives & Edited
by Robert Godwin

Volume 1

Special thanks to Buzz Aldrin
& Mark Kahn at NASA.

All rights reserved under article two of the Berne Copyright Convention (1971).
We acknowledge the financial support of the Government of Canada through the
Book Publishing Industry Development Program for our publishing activities.
Published by Apogee Books an imprint of Collector's Guide Publishing Inc., Box 62034, Burlington, Ontario, Canada, L7R 4K2
Printed and bound in Canada by Webcom Ltd of Toronto
Apollo 11 — The NASA Mission Reports
Edited by Robert Godwin
ISBN 1-896522-53-X
© 1999
All photos courtesy of NASA

In the last thirty years I have often reflected on the events that led up to the flight and total success of Apollo 11. The enormous efforts made by every single person who worked on the Apollo program should not be forgotten. Apollo was one of the greatest triumphs of cooperation in history.

The engineers who conceived and constructed the Apollo-Saturn space vehicle are the unsung heroes of our journey to the moon.

People often wonder what extraordinary character trait is required to become an astronaut. The simple truth is that we believed in the people around us and in ourselves. We trained hard and we knew that we were flying as good a vehicle as could be built. We were just the part of the team that had to fly those amazing machines.

During the flight I marveled at the performance of Saturn, Columbia, Eagle and the EVA back-packs. After only four manned flights the bugs had been worked out of the system but not without great sacrifice.

The tragedy of Apollo 1 with the loss of her crew was a sober reminder to all of us that space exploration is a dangerous business and that such successes often come with a high price.

In most histories of the space race the attention is focused on the pilots. We were the lucky ones, but we were part of a much larger team.

NASA was, and still is, an assemblage of talented people from all walks of life. Not just astronauts and test-pilots, but everyone from the theoreticians and engineers who built Apollo to the administrators and managers who organized it. From the safety officers and flight controllers who took care of us to the workers at the factories around America who put their hearts into every component.

We also shouldn't forget the contributions of the Congressional Representatives, Senators, and of course the Presidents who had the vision to initiate and see this extraordinary accomplishment through to its successful conclusion.

Those thousands of people around the United States who designed and planned and dreamed and schemed are the people who we are reminded of in this book. They are the ones who made it happen.

Dr Edwin "Buzz" Aldrin

Introduction

When the Saturn V carrying Apollo 11 left the launch pad in the Everglades of Cape Kennedy on July 16th 1969 it represented the zenith of humanity's scientific achievements. Robert Goddard, the early American rocketry pioneer, would have easily recognised the principles at play in Von Braun's behemoth as it slowly inched its way clear of the launch tower. Indeed, the ancient Chinese alchemists of a thousand years earlier would have seen fragments of their own designs.

The scientific titans of the Renaissance, such as Sir Isaac Newton, Galileo, Kepler and Copernicus would each have understood the celestial mechanics and inexorable physical rules which would guide the crew to their ultimate destination and then bring them home again.

Herbert George Wells and Jules Verne, two of the most imaginative minds to ever put pen to paper, might even have recognised some of their own ideas and both would have recognised the motivations behind the quest to put men on the moon. Not just the fulfillment of a noble goal, an inexorable next step — but also the continuing echoes of nationalism that both men so feared in their own century.

All of these great men of the past imagined a day when man might be able to reach out into the universe — not so much to tame it, but to live in harmony and understanding of it.

Neil Armstrong's proclamation of "...a giant leap for mankind." would have surely brought tears to the eyes of many of these visionaries but how many of them could have foreseen the technological requirements to make that giant leap?

Apollo 11 was the result of an unprecedented symbiosis of political will and technological capability.

President John Kennedy was the first leader in human history in a position to make such an audacious challenge. Even five or ten years earlier, all of the good intentions in the world would not have been able to push through such a project. Kennedy and his advisors were the first to recognise that they lived in an era when it *might* be possible technologically. It was also an era when it was politically expedient.

Throughout history progress and paranoia have often walked side-by-side. President Lyndon Johnson is ostensibly credited as saying he would "...not go to bed by the light of a communist moon!" The continued march of communism around the world had fanned the embers of resistance in a generation that had fought hard to maintain its freedoms on the harsh battlefields of World War II. Hot war had turned to Cold War and, in the wake of the McCarthy era, paranoia was still rampant.

When the rallying cry came, the people of America, responded with two voices. The political voice needed to establish the high ground, while the scientists and visionaries voiced the age-old dream. Could it be done?

While World War II was the defining moment of the 20th Century, it was also the most destructive. Out of that great conflagration came moments of utter horror and despair that we as a species will be striving to understand for centuries, but ironically it also gave birth to modern rocketry. If it wasn't for Peenemunde and the V-2, there could not have been Cape Kennedy and the Saturn V, at least not in the 1960's.

Parallels have been drawn between the scope of the Apollo project and the construction of the great pyramids of ancient Egypt. Kennedy as the benevolent Pharaoh, proclaiming that, "...it will be done". The appeal of a vast and non-belligerent goal. The marshaling of huge quantities of people and resources on a titanic scale. All of these things combined with humanity's latent need to explore and control our environment ultimately led to Apollo 11.

Looking back, with the benefit of hindsight, it is easy for many to question the legacy of project Apollo.
"Why did we spend so much money in space?"
 "What did we get out of it?"
 "Why didn't we stay?"
 "What was the point?"

These questions, and many others, resound again and again when people talk of Apollo. The answers are all out there, but, as is often the case, people don't want to hear them.

Nobody spent any money in space. There are no shops or factories in space. The money was spent here — on Earth. It created the world's wealthiest and most powerful and influential Aerospace industry. It bought jobs and it bought ideas. It bought infrastructure and technology. It bought medicine and it bought power and status. How might the political landscape look now had the Soviet Union been the winners of the so-called space race?

Apollo 11 brought the United States of America unparalleled prestige. Perhaps for the first and only time in history, all of mankind doffed their caps in a collective salute to a nation. Even the Soviets couldn't deny the moment. For a brief instant we were no longer Americans or Russians, or black or white, or Muslim or Christian — we were one family — human. We all participated and we all wanted to be in Armstrong and Aldrin's shoes. We were awe-inspired and we were proud, we were humbled and we were thrilled — and then — they came home.

We couldn't stay. Science fiction writer Robert Heinlein said it best in the title of his novel, "The Moon is a Harsh Mistress". Indeed, the moon is a hostile and dangerous place. Even the greatest minds of science fiction did not predict how utterly barren and lifeless a place it is. Project Apollo extinguished many fanciful notions of comfy vacations and hotels on the moon. Although at the end of the 20th century there is now compelling evidence of water in the shadows of the lunar poles, the lack of atmosphere, the extreme temperatures and the harsh radiation still add up to a daunting combination of obstacles.

The scientific results of the Apollo missions served to reinforce the arguments of the skeptics. What could be done on the moon that couldn't be done either here on Earth or in near-Earth orbit? It's easier to mine an asteroid than the moon. In fact it's probably easier to mine Mars than the moon. The current conventional-wisdom is that the only thing that the moon is any use for is as a base for a far-side observatory. Because the Earth is an incredibly noisy place, the far-side of the moon is a perfect place to build a large radio-telescope. As intriguing as this may be it's not very romantic. No, the people need adventure if they're going to put down their hard-earned cash. Staying on the moon in a subterranean bunker constantly in fear of de-pressurisation, solar flares and depletion of resources is not going to be much of an inducement to take a vacation. So what was gained?

Knowledge and experience.

Although robot probes can answer many questions for scientists, the general public need heroes and they need to participate. People need to touch, poke, sniff and prod. As a species we are innately suspicious and curious. If we don't get to use our own senses we are never quite sure of anything.

We all lived vicariously through the Apollo astronauts. After Apollo 11 our moon would never be the same. We had seen pictures from the Ranger and Surveyor and Luna probes but no one had actually picked up a piece of moon and held it in their hand. Until that moment we continued to doubt. Apollo 11 gave us the knowledge and confidence to move on.

At the dawn of a new millennium planning is under way to send people to Mars. The best theoreticians of today are dreaming up ways for us to take the next giant leap. Lessons learned from Apollo are being used every day to discuss how we can build those romantic hotels on Mars and in orbit around the Earth. All we need now is the political will to build on the Apollo legacy. The knowledge, technology, resources and manpower are available.

Trouble is — you can never find a Pharaoh when you need one.

Robert Godwin (Editor)

Contents

Apollo 11 Press Kit

Pre-Flight Mission Operation Report

LIST OF FIGURES

LIST OF TABLES

Post-Flight Director's Mission Operation Report

Post-Flight Crew Press Conference

LUNAR LANDING MISSION PRESS KIT

NATIONAL AERONAUTICS AND SPACE ADMINISTRATION

FOR RELEASE: SUNDAY July 6, 1969

RELEASE NO: 69-83K

PROJECT: APOLLO 11

(To be launched no earlier than July 16)

APOLLO 11

The United States will launch a three-man spacecraft toward the Moon on July 16 with the goal of landing two astronaut explorers on the lunar surface four days later.

If the mission — called Apollo 11 — is successful, man will accomplish his long-time dream of walking on another celestial body.

The first astronaut on the Moon's surface will be 38-year-old Neil A. Armstrong of Wapakoneta, Ohio, and his initial act will be to unveil a plaque whose message symbolizes the nature of the journey.

Affixed to the leg of the lunar landing vehicle, the plaque is signed by President Nixon, Armstrong and his Apollo 11 companions, Michael Collins and Edwin E. Aldrin, Jr.

It bears a map of the Earth and this inscription:

HERE MEN FROM THE PLANET EARTH FIRST SET FOOT UPON THE MOON JULY 1969 A.D.

WE CAME IN PEACE FOR ALL MANKIND

The plaque is fastened to the descent stage of the lunar module and thus becomes a permanent artifact on the lunar surface.

Later Armstrong and Aldrin will implant an American flag on the surface of the Moon.

The Apollo 11 crew will also carry to the Moon and return two large American flags, flags of the 50 states, District of Columbia and U.S. Territories, flags of other nations and that of the United Nations Organization.

During their 22-hour stay on the lunar surface, Armstrong and Aldrin will spend up to 2 hours and 40 minutes outside the lunar module, also gathering samples of lunar surface material and deploying scientific experiments which will transmit back to Earth valuable data on the lunar environment.

Apollo 11 is scheduled for launch at 9:32 a.m. EDT July 16 from the National Aeronautics and Space Administration's Kennedy Space Center Launch Complex 39-A. The mission will be the fifth manned Apollo flight and the third to the Moon.

The prime mission objective of Apollo 11 is stated simply: "Perform a manned lunar landing and return". Successful fulfillment of this objective will meet a national goal of this decade, as set by President Kennedy May 25, 1961.

Apollo 11 Commander Armstrong and Command Module Pilot Collins 38, and Lunar Module Pilot Aldrin, 39, will each be making his second space flight. Armstrong was Gemini 8 commander, and backup Apollo 8 commander; Collins was Gemini 10 pilot and was command module pilot on the Apollo 8 crew until spinal surgery forced him to leave the crew for recuperation; and Aldrin was Gemini 12 pilot and Apollo 8 backup lunar module pilot. Armstrong is a civilian, Collins a USAF lieutenant colonel and Aldrin a USAF colonel.

Apollo 11 backup crewmen are Commander James A. Lovell, Command Module Pilot William A. Anders, both of whom were on the Apollo 8 first lunar orbit mission crew, and Lunar Module Pilot Fred W. Haise.

The backup crew functions in three significant categories. They help the prime crew with mission preparation and hardware checkout activities. They receive nearly complete mission training which becomes a valuable foundation for later assignment as a prime crew and finally, should the prime crew become unavailable, they are prepared to fly as prime crew on schedule up until the last few weeks at which time full duplicate training becomes too costly and time consuming to be practical.

Apollo 11, after launch from Launch Complex 39-A, will begin the three-day voyage to the Moon about two and a half hours after the spacecraft is inserted into a 100-nautical mile circular Earth parking orbit. The Saturn V launch vehicle third stage will restart to inject Apollo 11 into a translunar trajectory as the vehicle passes over the Pacific midway through the second Earth parking orbit.

The "go" for translunar injection will follow a complete checkout of the space vehicle's readiness to be committed for injection. About a half hour after translunar injection (TLI), the command/ service module will separate from the Saturn third stage, turn around and dock with the lunar module nested in the spacecraft LM adapter. Spring-loaded lunar module holddowns will be released to eject the docked spacecraft from the adapter.

APOLLO 11 —— Launch And Translunar Injection

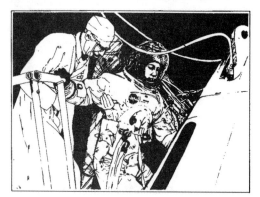
Astronaut Insertion

Check Of Systems

Saturn Staging

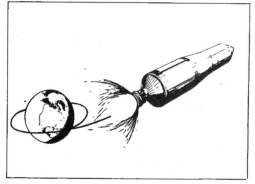
Translunar Injection

Later, leftover liquid propellant in the Saturn third stage will be vented through the engine bell to place the stage into a "slingshot" trajectory to miss the Moon and go into solar orbit.

During the translunar coast, Apollo 11 will be in the passive thermal control mode in which the spacecraft rotates slowly about one of its axes to stabilize thermal response to solar heating. Four midcourse correction maneuvers are possible during translunar coast and will be planned in real time to adjust the trajectory.

Apollo 11 will first be inserted into a 60-by-170-nautical mile elliptical lunar orbit, which two revolutions later will be adjusted to a near-circular 54 x 66 nm. Both lunar orbit insertion burns (LOI), using the spacecraft's 20,500-pound-thrust service propulsion system, will be made when Apollo 11 is behind the Moon and out of "sight" of Manned Space Flight Network stations.

Some 21 hours after entering lunar orbit, Armstrong and Aldrin will man and check out the lunar module for the descent to the surface. The LM descent propulsion system will place the LM in an elliptical orbit with a pericynthion, or low point above the Moon, of 50,000 feet, from which the actual descent and touchdown will be made.

APOLLO 11 —— Translunar Flight

Transposition Maneuver

Extraction Of Lunar Module

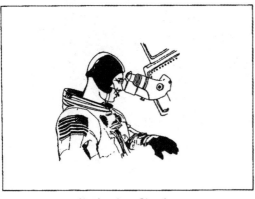

Navigation Check

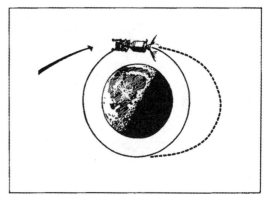

Lunar Orbit Insertion

After touchdown, the landing crew will first ready the lunar module for immediate ascent and then take a brief rest before depressurizing the cabin for two-man EVA about 10 hours after touchdown. Armstrong will step onto the lunar surface first, followed by Aldrin some 40 minutes later.

During their two hours and 40 minutes on the surface, Armstrong and Aldrin will gather geologic samples for return to Earth in sealed sample return containers and set up two scientific experiments for returning Moon data to Earth long after the mission is complete.

One experiment measures moonquakes and meteoroid impacts on the lunar surface, while the other experiment is a sophisticated reflector that will mirror laser beams back to points on Earth to aid in expanding scientific knowledge both of this planet and of the Moon.

The lunar module's descent stage will serve as a launching pad for the crew cabin as the 3,500-pound-thrust ascent engine propels the LM ascent stage back into lunar orbit for rendezvous with Collins in the command/service module— orbiting 60 miles above the Moon.

Four basic maneuvers, all performed by the LM crew using the spacecraft's small maneuvering and attitude thrusters, will bring the LM and the command module together for docking about three and a half hours after liftoff from the Moon.

The boost out of lunar orbit for the return journey is planned for about 135 hours after Earth liftoff and after the LM ascent stage has been jettisoned and lunar samples and film stowed aboard the command module. An optional plan provides for a 12-hour delay in the transearth injection burn to allow the crew more rest after a long hard day's work on the lunar surface and flying the rendezvous. The total mission time to splashdown would remain about the same, since the transearth injection burn would impart a higher velocity to bring the spacecraft back to the mid-Pacific recovery line at about the same time.

APOLLO 11 — Descent To Lunar Surface

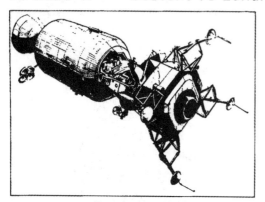

Transfer To LM

Separation Of LM From CSM

Landing On Moon

First Step On Moon

APOLLO 11 — Lunar Surface Activities

Commander On Moon

Contingency Sample

Documented Sample Collection

Sample Collecting

APOLLO 11 —— Lunar Surface Activities

Experiment Placements

TV Camera

Alignment Of Passive
Seismometer

Bulk Sample Collection

APOLLO 11 —— Lunar Ascent And Rendezvous

Return To Spacecraft

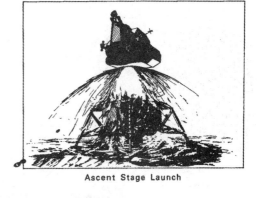

Ascent Stage Launch

Rendezvous And Docking

LM Jettison

The rendezvous sequence to be flown on Apollo 11 has twice been flown with the Apollo spacecraft — once in Earth orbit on Apollo 9 and once in lunar orbit with Apollo 10. The Apollo 10 mission duplicated, except for the actual landing, all aspects of the Apollo 11 timeline.

APOLLO 11 — Transearth Injection And Recovery

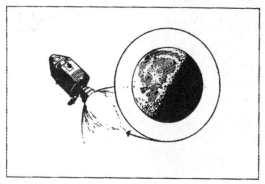

Transearth Injection

CM/SM Separation

Reentry

Recovery

Splashdown

During the transearth coast period, Apollo 11 will again control solar heat loads by using the passive thermal control "barbeque" technique. Three transearth midcourse corrections are possible and will be planned in real time to adjust the Earth entry corridor.

Apollo 11 will enter the Earth's atmosphere (400,000 feet) at 195 hours and five minutes after launch at 36,194 feet per second. Command module touchdown will be 1285 nautical miles downrange from entry at 10.6 degrees north latitude by 172.4 west longitude at 195 hours, 19 minutes after Earth launch 12:46 p.m. EDT July 24. The touchdown point is about 1040 nautical miles southwest of Honolulu, Hawaii.
(END OF GENERAL RELEASE; BACKGROUND INFORMATION FOLLOWS)

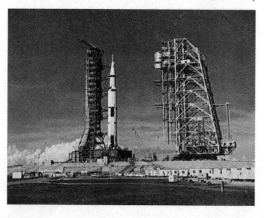

Official Apollo 11 Insignia

This photograph not for release
before Saturday, July 5, 1969

FLIGHT PROFILE

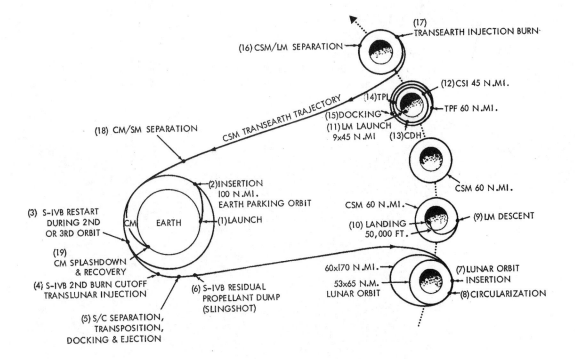

APOLLO 11 COUNTDOWN

The clock for the Apollo 11 countdown will start at T-28 hours, with a six hour built-in-hold planned at T-9 hours, prior to launch vehicle propellant loading.

The countdown is preceded by a pre-count operation that begins some 5 days before launch. During this period the tasks include mechanical buildup of both the command/service module and LM, fuel cell activation and servicing and loading of the super critical helium aboard the LM descent stage.

Following are some of the highlights of the final count:

T-28 hrs.	Official countdown starts
T-27 hrs. 30 mins.	Install launch vehicle flight batteries (to 23 hrs. 30 mins.)
	LM stowage and cabin closeout (to 15 hrs.)
T-21 hrs.	Top off LM super critical helium (to 19 hrs.)
T-16 hrs.	Launch vehicle range safety checks (to 15 hrs.)
T-11 hrs. 30 mins.	Install launch vehicle destruct devices (to 10 hrs. 45 mins.)
	Command/service module pre-ingress operations
T-10 hrs.	Start mobile service structure move to park site
T-9 hrs.	Start six hour built-in-hold
T-9 hrs. counting	Clear blast area for propellant loading
T-8 hrs. 30 mins.	Astronaut backup crew to spacecraft for prelaunch checks
T-8 hrs. 15 mins.	Launch Vehicle propellant loading three stages (liquid oxygen in first stage)
	liquid oxygen and liquid hydrogen in second, third stages
	Continues thru T-3 hrs: 38 mins.
T-5 hrs. 17 mins.	Flight crew alerted
T-5 hrs. 02 mins.	Medical examination
T-4 hrs. 32 mins.	Breakfast
T-3 hrs. 57 mins.	Don space suits
T-3 hrs. 07 mins.	Depart Manned Spacecraft Operations Building for LC 39 via crew transfer van
T-2 hrs, 55 mins.	Arrive at LC-39
T-2 hrs. 40 mins.	Start flight crew ingress
T-1 hr. 55 mins.	Mission Control Center-Houston/ spacecraft command checks
T-1 hr. 50 mins.	Abort advisory system checks
T-1 hr. 46 mins.	Space vehicle Emergency Detection System (EDS) test
T-43 mins.	Retrack Apollo access arm to standby position (12 degrees)
T-42 mins.	Arm launch escape system
T-40 mins,	Final launch vehicle range safety checks (to 35 mins.)
T-30 mins.	Launch vehicle power transfer test LM switch over to internal power
T-20 mins. to T-10 mins.	Shutdown LM operational instrumentation
T-15 mins.	Spacecraft to internal power
T-6 mins.	Space vehicle final status checks
T-5 mins. 30 sec.	Arm destruct system
T-5 mins.	Apollo access arm fully retracted
T-3 mins. 10 sec.	Initiate firing command (automatic sequencer)
T-50 sec.	Launch vehicle transfer to internal power
T-8.9 sec.	Ignition sequence start
T-2 sec.	All engines running
T-0	Liftoff

*Note: Some changes in the above countdown are possible as a result of experience gained in the Countdown Demonstration Test (CDDT) which occurs about 10 days before launch.

LAUNCH EVENTS

Time Hrs Min Sec	Event	Altitude Feet	Velocity Ft/Sec	Range Nau Mi
00 00 00	First Motion	182.7	1,340.67	0.0
00 01 21.0	Maximum Dynamic Pressure	43,365	2,636.7	2.7
00 02 15	S-IC Center Engine Cutoff	145,600	6,504.5	24.9
00 02 40.8	S-IC Outboard Engines Cutoff	217,655	9,030.6	49.6
00 02 41.6	S-IC/S-II Separation	219,984	9,064.5	50.2
00 02 43.2	S-II Ignition	221,881	9,059.1	51.3
00 03 11.5	S-II Aft Interstage Jettison	301,266	9,469.0	87.0
00 03 17.2	LET Jettison	315,001	9,777.6	94.3
00 07 39.8	S-II Center Engine Cutoff	588,152	18,761.7	600.0
00 09 11.4	S-II Outboard Engines Cutoff	609,759	22,746.8	885.0
00 09 12.3	S-II/S-IVB Separation	609,982	22,756.7	887.99
00 09 15.4	S-IVB Ignition	610,014	22,756.7	888.42
00 11 40.1	S-IVB First Cutoff	617,957	25,562.4	1425.2
00 11 50.1	Parking Orbit Insertion	617,735	25,567.9	1463.9
02 44 14.8	S-IVB Reignition	650,558	25,554.0	3481.9
02 50 03.1	S-IVB Second Cutoff	1,058,809	35,562.9	2633.6
02 50 13.1	Translunar Injection	1,103,215	35,538.5	2605.0

APOLLO 11 MISSION EVENTS

Event	GET Hrs:min:sec	Date/EDT	Vel.Change feet/sec	Purpose and resultant orbit
Earth orbit insertion	00:11:50	16th 9:44 a	25,567	Insertion into 100 nm circular earth parking orbit
Translunar injection (S-IVB engine ignition)	02:44:15	16th 12:16 p	9,965	Injection into free-return trans lunar trajectory with 60 nm pericynthion
CSM separation, docking	03:20:00	16th 12:52 p		Hard-mating of CSM and LM
Ejection from SLA	04:10:00	16th 1:42 p	1	Separates CSM-LM from S-IVB-SLA
SPS Evasive maneuver	04:39:37	16th 2:12 p	19.7	Provides separation prior to S-IVB propellant dump and "slingshot" maneuver
Midcourse correction #1	TLI+9 hrs	16th 9:16 p	*0	*These midcourse corrections have a nominal velocity change of 0 fps,
Midcourse correction #2	TLI+24 hrs	17th 12:16 p	0	but will be calculated in real time to correct TLI dispersions.
Midcourse correction #3	LOI-22 hrs	18th 3:26 p	0	
Midcourse correction #4	LOI-5 hrs	19th 8:26 a	0	
Lunar orbit insertion No. 1	75:54:28	19th 1:26 p	-2924	Inserts Apollo 11 into 60 x 170 nmelliptical lunar orbit

Event	GET Hrs:min:sec	Date/EDT	Vel.Change feet/sec	Purpose and resultant orbit
Lunar orbit insertion No. 2	80:09:30	19th 5:42 p	-157.8	Changes lunar parking orbit to 54 x 66 nm
CSM-LM undocking, separation (SM RCS)	100:09:50 100:39:50	20th 1:42 p 20th 2:12 p	— 2.5	Establishes equiperiod orbit for 2.2 nm separation for DOI maneuver
Descent orbit insertion (DPS)	101:38:48	20th 3:12 p	-74.2	Lowers LM pericynthion to 8 nm (8 x 60)
LM powered descent initiation (DPS)	102:35:13	20th 4:08 p	-6761	Three-phase maneuver to brake LM out of transfer orbit, vertical descent and touchdown on lunar surface
LM touchdown on lunar surface	102:47:11	20th 4:19 p		Lunar exploration
Depressurization for lunar surface EVA	112:30	21st 2:02 a		
Repressurize LM after EVA	115:10	21st 4:42 a		
LM ascent and orbit insertion	124:23:21	21st 1:55 p	6055	Boosts ascent stage into 9 x 45 lunar orbit for rendezvous with CSM
LM RCS concentric sequence initiate (CSI) burn	125:21:20	21st 2:53 p	49.4	Raises LM perilune to 44.7 nm, adjusts orbital shape for rendezvous sequence (45.5 x 44.2)
LM RCS constant delta height (CDH) burn	126:19:40	21st 3:52 p	4.5	Radially downward burn adjusts LM orbit to constant 15 nm below CSM
LM RCS terminal phase initiate (TPI) burn	126:58:26	21st 4:30 p	24.6	LM thrusts along line of sight toward CSM, midcourse and braking maneuvers as necessary
Rendezvous (TPF)	127:43:54	21st 5:15 p	-4.7	Completes rendezvous sequence (59.5 x 59.0)
Docking	128:00:00	21st 5:32 p		Commander and LM pilot transfer back to CSM
LM jettison, separation (SM RCS)	131:53:05	21st 9:25 p	-1	Prevents recontact of CSM with LM ascent stage during remainder of lunar orbit
Transearth injection (TEI) SPS	135:24:34	22nd 00:57 a	3293	Inject CSM into 59.6-hour trans earth trajectory

Event	GET Hrs:min:sec	Date/EDT	Vel.Change feet/sec	Purpose and resultant orbit
Midcourse correction No. 5	TEI+15 hrs	22nd 3:57 p	0	Transearth midcourse corrections will be computed in real time for entry corridor
Midcourse correction No. 6	EI -15 hrs	23rd 9:37 p	0	control and recovery area weather avoidance.
Midcourse correction No. 7	EI -3 hrs	24th 9:37 a	0	
CM/SM separation	194:50:04	24th 12:22 p	—	Command module oriented for entry
Entry interface (400,000 feet)	195:05:04	24th 12:37 p	—	Command module enters earth's sensible atmosphere at 36,194 fps
Touchdown	195:19:05	24th 12:51 p	—	Landing 1285 nm downrange from entry, 10.6 north latitude by 172.4 west longitude.

MISSION TRAJECTORY AND MANEUVER DESCRIPTION

Information presented herein is based upon a July 16 launch and is subject to change prior to the mission or in real time during the mission to meet changing conditions.

Launch

Apollo 11 will be launched from Kennedy Space Center Launch Complex 39A on a launch azimuth that can vary from 72 degrees to 106 degrees, depending upon the time of day of launch. The azimuth changes with time of day to permit a fuel optimum injection from Earth parking orbit into a free-return circumlunar trajectory. Other factors influencing the launch windows are a daylight launch and proper Sun angles on the lunar landing sites.

The planned Apollo 11 launch date of July 16 will call for liftoff at 9:32 a.m. EDT on a launch azimuth of 72 degrees. The 7.6-million-pound thrust Saturn V first stage boosts the space vehicle to an altitude of 36.3 nm at 50.6 nm downrange and increases the vehicle's velocity to 9030.6 fps in 2 minutes 40.8 seconds of powered flight. First stage thrust builds to 9,088,419 pounds before center engine shutdown. Following out-board engine shutdown, the first stage separates and falls into the Atlantic Ocean about 340 nm downrange (30.3 degrees North latitude and 73.5 degrees West longitude) some 9 minutes after liftoff.

The 1-million-pound thrust second stage (S-II) carries the space vehicle to an altitude of 101.4 nm and a distance of 885 nm downrange. Before engine burnout, the vehicle will be moving at a speed of 22,746.8 fps. The outer J-2 engines will burn 6 minutes 29 seconds during this powered phase, but the center engine will be cut off at 4 minutes 56 seconds after S-II ignition.

At outboard engine cutoff, the S-II separates and, following a ballistic trajectory, plunges into the Atlantic Ocean about 2,300 nm downrange from the Kennedy Space Center (31 degrees North latitude and 33.6 degrees West longitude) some 20 minutes after liftoff.

The first burn of the Saturn V third stage (S-IVB) occurs immediately after S-II stage separation. It will last long enough (145 seconds) to insert the space vehicle into a circular Earth parking orbit beginning at about 4,818 nm downrange. Velocity at Earth orbital insertion will be 25,567 fps at 11 minutes 50 seconds ground elapsed time (GET). Inclination will be 32.6 degrees.

LAUNCH WINDOW SUMMARY

		16	18	21	
	LAUNCH DATE	16	18	21	
JULY	LAUNCH WINDOW, E.D.T.	9:32-13:54	9:38-14:02	10:09-14:39	
16-21	SITE/PROFILE	2/FR	3/FR	5/HYB	
	SUN ELEVATION ANGLE	9.9-12.6	8.3-11.0	6.3-9.0	
	MISSION TIME, DAYS:HOURS	8:3	8:5	8:8	
	SPS RESERVES, FPS	1700	1550	1750	
	LAUNCH DATE	14	16		20
AUGUST	LAUNCH WINDOW, E.D.T.	7:51-12:15	8:04-12:31		10:05-14:47
14-20	SITE/PROFILE	2/HYB	3/HYB		5/HYB
	SUN ELEVATION ANGLE	6.2-8.9	6.2-8.9		9.0-12.0
	MISSION TIME, DAYS:HOURS	8:5	8:7		8:8
	SPS RESERVES, FPS	1600	1750		1300
	LAUNCH DATE	13	15	18	
SEP	LAUNCH WINDOW, E.D.T.	6:17-10:45	7:04-11:39	11:31-16:14	
13-18	SITE/PROFILE	2/HYB	3/HYB	5/HYB	
	SUN ELEVATION ANGLE	6.8-9.6	6.3-91.2	6.8-9.7	
	MISSION TIME, DAYS:HOURS	8:7	8:8	8:6	
	SPS RESERVES, FPS	1600	1500	1050	

The crew will have a backup to launch vehicle guidance during powered flight. If the Saturn instrument unit inertial platform fails, the crew can switch guidance to the command module systems for first-stage powered flight automatic control. Second and third stage backup guidance is through manual takeover in which crew hand controller inputs are fed through the command module computer to the Saturn instrument unit.

Earth Parking Orbit (EPO)

Apollo 11 will remain in Earth parking orbit for one-and-one half revolutions after insertion and will hold a local horizontal attitude during the entire period. The crew will perform spacecraft systems checks in preparation for the translunar injection (TLI) burn. The final "go" for the TLI burn will be given to the crew through the Carnarvon, Australia, Manned Space Flight Network station.

Translunar Injection (TLI)

Midway through the second revolution in Earth parking orbit, the S-IVB thirdstage engine will restart at 2:44:15 GET over the mid-Pacific just south of the equator to inject Apollo 11 toward the Moon. The velocity will increase from 25,567 fps to 35,533 fps at TLI cutoff — a velocity increase of 9971 fps. The TLI burn is targeted for about 6 fps overspeed to compensate for the later SPS evasive maneuver after LM extraction. TLI will place Apollo 11 on a free-return circumlunar trajectory from which midcourse corrections if necessary could be made with the SM RCS thrusters. Entry from a free-return trajectory would be at 10:37 a.m. EDT July 22 at 14.9 degrees south latitude by 174.9 east longitude after a flight time of 145 hrs 04 min.

MISSION DURATIONS

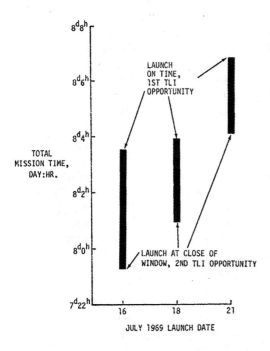

TOTAL
MISSION TIME,
DAY:HR.

LAUNCH
ON TIME,
1ST TLI
OPPORTUNITY

LAUNCH AT CLOSE OF
WINDOW, 2ND TLI OPPORTUNITY

JULY 1969 LAUNCH DATE

Transposition, Docking and Ejection (TD&E)

At about three hours after liftoff and 25 minutes after the TLI burn, the Apollo 11 crew will separate the command/service module from the spacecraft lunar module adapter (SLA), thrust out away from the S-IVB, turn around and move back in for docking with the lunar module. Docking should take place at about three hours and 21 minutes GET, and after the crew confirms all docking latches solidly engaged, they will connect the CSM-to-LM umbilicals and pressurize the LM with the command module surge tank. At about 4:09 GET, the spacecraft will be ejected from the spacecraft LM adapter by spring devices at the four LM landing gear "knee" attach points. The ejection springs will impart about one fps velocity to the spacecraft. A 19.7 fps service propulsion system (SPS) evasive maneuver in plane at 4.39 GET will separate the spacecraft to a safe distance for the S-IVB "slingshot" maneuver in which residual launch vehicle liquid propellants will be dumped through the J-2 engine bell to propel the stage into a trajectory passing behind the Moon's trailing edge and on into solar orbit.

HYBRID LUNAR PROFILE

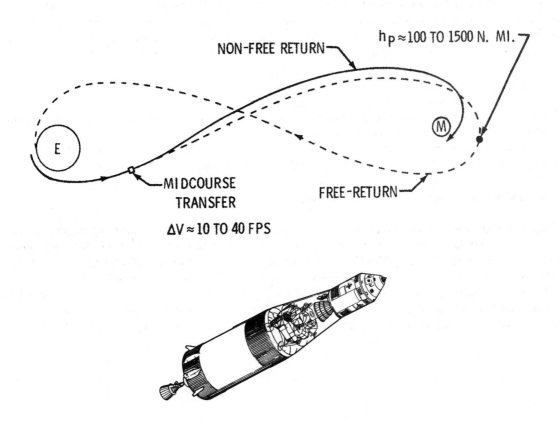

$h_p \approx 100$ TO 1500 N. MI.

NON-FREE RETURN

MIDCOURSE
TRANSFER

FREE-RETURN

$\Delta V \approx 10$ TO 40 FPS

VEHICLE EARTH PARKING ORBIT CONFIGURATION
(SATURN V THIRD STAGE AND INSTRUMENT UNIT, APOLLO SPACECRAFT)

POST TLI TIMELINE

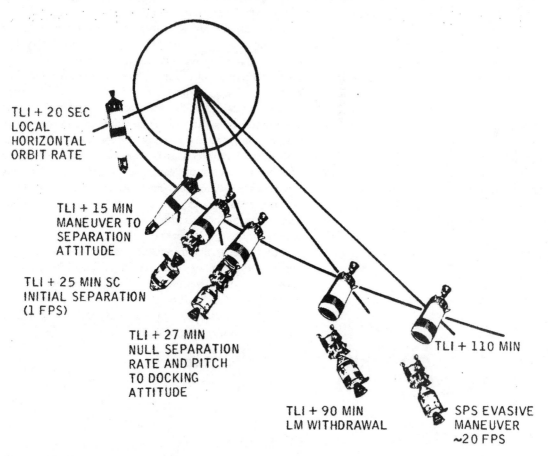

TLI + 20 SEC
LOCAL
HORIZONTAL
ORBIT RATE

TLI + 15 MIN
MANEUVER TO
SEPARATION
ATTITUDE

TLI + 25 MIN SC
INITIAL SEPARATION
(1 FPS)

TLI + 27 MIN
NULL SEPARATION
RATE AND PITCH
TO DOCKING
ATTITUDE

TLI + 90 MIN
LM WITHDRAWAL

TLI + 110 MIN

SPS EVASIVE
MANEUVER
~20 FPS

Translunar Coast

Up to four Midcourse correction burns are planned during the translunar coast phase, depending upon the accuracy of the trajectory resulting from the TLI maneuver. If required, the midcourse correction burns are planned at TLI +9 hours, TLI +24 hours, lunar orbit insertion (LOI) -22 hours and LOI -5 hours.

During coast periods between midcourse corrections, the spacecraft will be in the passive thermal control (PTC) or "barbecue" mode in which the spacecraft will rotate slowly about one axis to stabilize spacecraft thermal response to the continuous solar exposure.

Lunar Orbit Insertion (LOI)

The first of two lunar orbit insertion burns will be made at 75:54:28 GET at an altitude of about 80 nm above the Moon. LOI-1 will have a nominal retrograde velocity change of 2,924 fps and will insert Apollo 11 into a 60 x 170 nm elliptical lunar orbit. LOI-2 two orbits later at 80:09:30 GET will adjust the orbit to a 54 x 65 nm orbit, which because of perturbations of the lunar gravitational potential, will become circular at 60 nm at the time of rendezvous with the LM. The burn will be 157.8 fps retrograde. Both LOI maneuvers will be with the SPS engine near pericynthion when the spacecraft is behind the Moon and out of contact with MSFN stations. After LOI-2 (circularization), the lunar module pilot will enter the lunar module for a brief checkout and return to the command module.

Lunar Module Descent, Lunar Landing

The lunar module will be manned and checked out for undocking and subsequent landing on the lunar surface at Apollo site 2. Undocking will take place at 100:09:50 GET prior to the MSFN acquisition of signal. A radially downward service module RCS burn of 2.5 fps will place the CSM on an equiperiod orbit with a maximum separation of 2.2 nm one half revolution after the separation maneuver. At this point, on lunar farside, the descent orbit insertion burn (DOI) will be made with the lunar module descent engine firing retrograde 74.2 fps at 101:38:48 GET. The burn will start at 10 per cent throttle for 15 seconds and the remainder at 40 per cent throttle.

The DOI maneuver lowers LM pericynthion to 50,000 feet at a point about 14 degrees uprange of landing site 2.

A three-phase powered descent initiation (PDI) maneuver begins at pericynthion at 102:53:13 GET using the LM descent engine to brake the vehicle out of the descent transfer orbit. The guidance-controlled PDI maneuver starts about 260 nm prior to touchdown, and is in retrograde attitude to reduce velocity to essentially zero at the time vertical descent begins. Spacecraft attitudes range from windows down at the start of PDI, to windows up as the spacecraft reaches 45,000 feet above the lunar surface and LM landing radar data can be integrated by the LM guidance computer. The braking phase ends at about 7,000 feet above the surface and the spacecraft is rotated to an upright windows-forward attitude. The start of the approach phase is called high gate, and the start of the landing phase at 500 feet is called low gate.

Both the approach phase and landing phase allow pilot takeover from guidance control as well as visual evaluation of the landing site. The final vertical descent to touchdown begins at about 150 feet when all forward velocity is nulled out. Vertical descent rate will be three fps. Touchdown will take place at 102:47:11 GET.

LUNAR ORBIT INSERTION

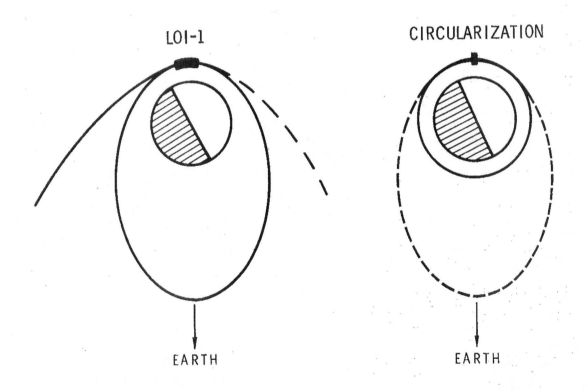

LOI-1

CIRCULARIZATION

EARTH   EARTH

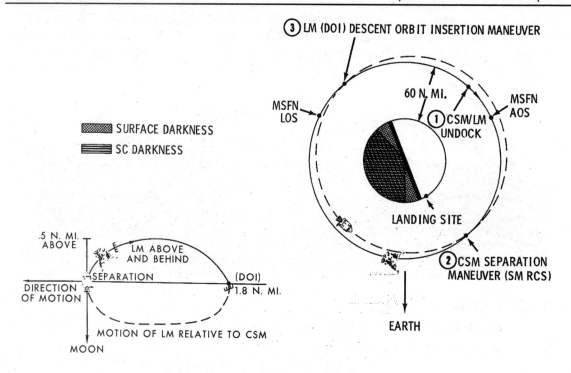

CSM/LM SEPARATION MANEUVER

LUNAR MODULE DESCENT

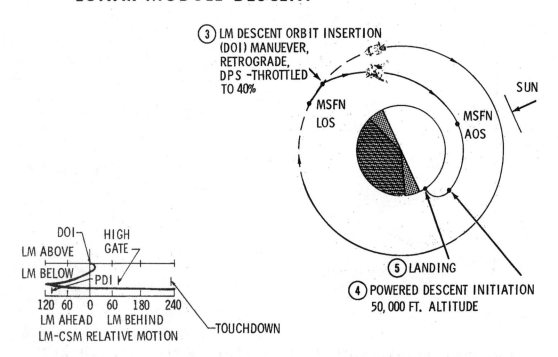

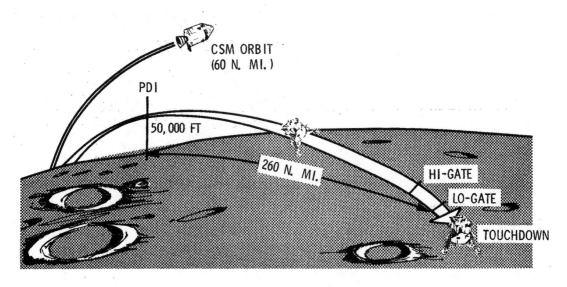

DESIGN CRITERIA

- BRAKING PHASE (PDI TO HI-GATE) - EFFICIENT REDUCTION OF ORBITAL VELOCITY
- FINAL APPROACH PHASE (HI-GATE TO LO-GATE) - CREW VISIBILITY (SAFETY OF FLIGHT AND SITE ASSESSMENT)
- LANDING PHASE (LO-GATE TO TOUCHDOWN) - MANUAL CONTROL TAKEOVER

PERATIONAL PHASES OF POWERED DESCENT

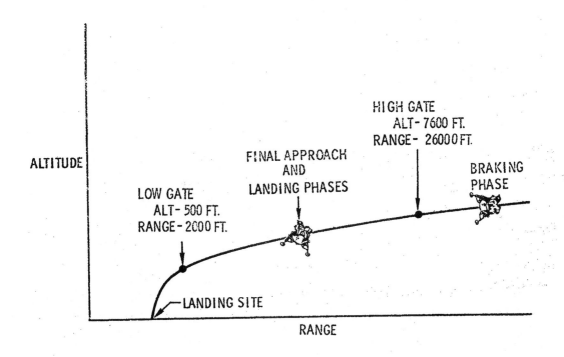

TARGET SEQUENCE FOR AUTOMATIC GUIDANCE

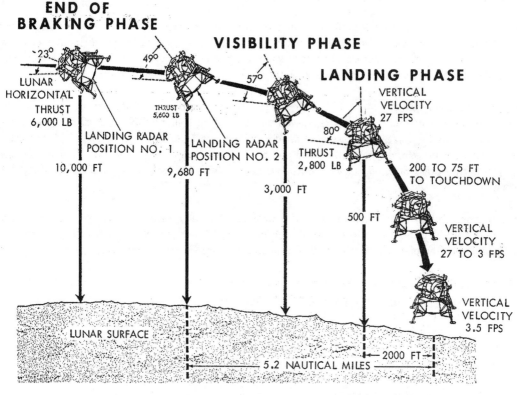

NOMINAL DESCENT TRAJECTORY
FROM HIGH GATE TO TOUCHDOWN

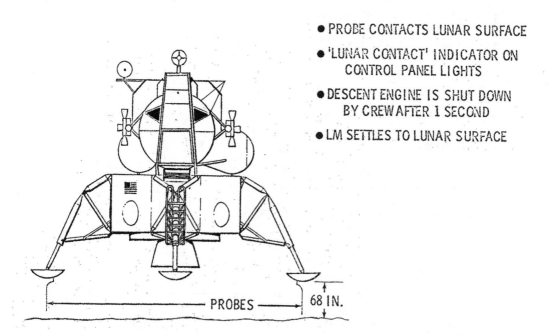

- PROBE CONTACTS LUNAR SURFACE
- 'LUNAR CONTACT' INDICATOR ON CONTROL PANEL LIGHTS
- DESCENT ENGINE IS SHUT DOWN BY CREW AFTER 1 SECOND
- LM SETTLES TO LUNAR SURFACE

LUNAR CONTACT SEQUENCE

Lunar Surface Extravehicular Activity (EVA)

Armstrong and Aldrin will spend about 22 hours on the lunar surface after lunar module touchdown at 102:47:11 GET. Following extensive checkout of LM systems and preparations for contingency ascent staging, the LM crew will eat and rest before depressurizing the LM for lunar surface EVA. Both crewmen will don portable life support system (PLSS) backpacks with oxygen purge system units (OPS) attached.

LM depressurization is scheduled for 112:30 GET with the commander being the first to egress the LM and step onto the lunar surface. His movements will be recorded on still and motion picture film by the lunar module pilot and by TV deployed by the commander prior to descending the ladder. The LM pilot will leave the LM about 25 minutes after the commander and both crewmen will collect samples of lunar material and deploy the Early Apollo Scientific Experiments Package (EASEP) and the solar wind composition (SWC) experiment.

The commander, shortly after setting foot on the lunar surface, will collect a contingency sample of surface material and place it in his suit pocket. Later both crewmen will collect as much as 130 pounds of loose materials and core samples which will be stowed in air-tight sample return containers for return to Earth.

Prior to sealing the SRC, the SWC experiment, which measures the elemental and isotopic constituents of the noble (inert) gases in the solar wind, is rolled up and placed in the container for return to Earth for analysis. Principal experimenter is Dr. Johannes Geiss, University of Bern, Switzerland.

The crew will photograph the landing site terrain and inspect the LM during the EVA. They can range out to about 100 feet from the LM.

After both crewmen have ingressed the LM and have connected to the cabin suit circuit, they will doff the PLSS backpacks and jettison them along with other gear no longer needed, through the LM front hatch onto the lunar surface.

The LM cabin will be repressurized about 2 hrs. 40 min. after EVA initiation to permit transfer by the crew to the LM life support systems. The LM will then be depressurized to jettison unnecessary equipment to the lunar surface and be repressurized. The crew will have a meal and rest period before preparing for ascent into lunar orbit and rendezvousing with the CSM.

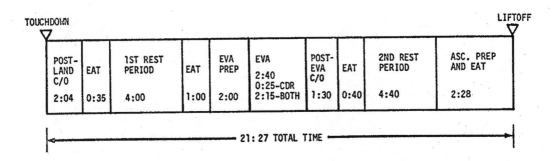

LUNAR SURFACE ACTIVITY SCHEDULE

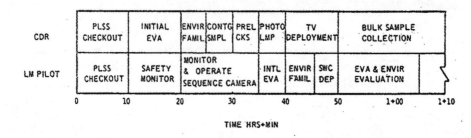

CDR	PLSS CHECKOUT	INITIAL EVA	ENVIR FAMIL	CONTG SMPL	PREL CKS	PHOTO LMP	TV DEPLOYMENT	BULK SAMPLE COLLECTION
LM PILOT	PLSS CHECKOUT	SAFETY MONITOR	MONITOR & OPERATE SEQUENCE CAMERA		INTL EVA	ENVIR FAMIL	SWC DEP	EVA & ENVIR EVALUATION

0   10   20   30   40   50   1+00   1+10

TIME HRS+MIN

CDR	LM INSPECTION	EASEP DEPLOYMENT	DOCUMENTED SAMPLE COLLECTION	REST PHOTO LMP PREPARE AND TRANSFER SRC'S	TERMINATE EVA
LM PILOT	LM INSPECTION	EASEP DEPLOYMENT	DOCUMENTED SAMPLE COLLECTION	TERMINATE EVA RECEIVE SRC'S	

1+10   1+20   1+30   1+40   1+50   2+00   2+10   2+20   2+30   2+40

TIME HRS+MIN

NOMINAL EVA TIMELINE

VIEW THRU OPTICAL CENTER OF TV LENS IN DIRECTION OF "Z"-PLANE

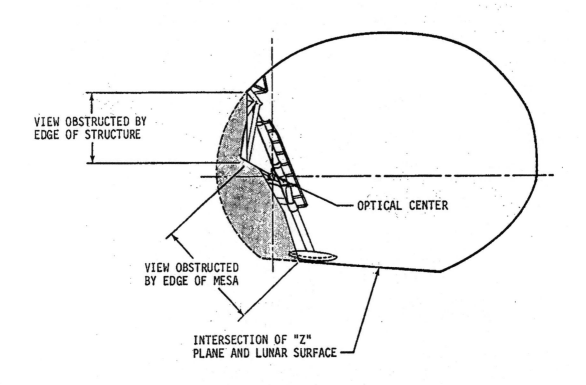

VIEW OBSTRUCTED BY EDGE OF STRUCTURE

OPTICAL CENTER

VIEW OBSTRUCTED BY EDGE OF MESA

INTERSECTION OF "Z" PLANE AND LUNAR SURFACE

LUNAR SURFACE PHASE

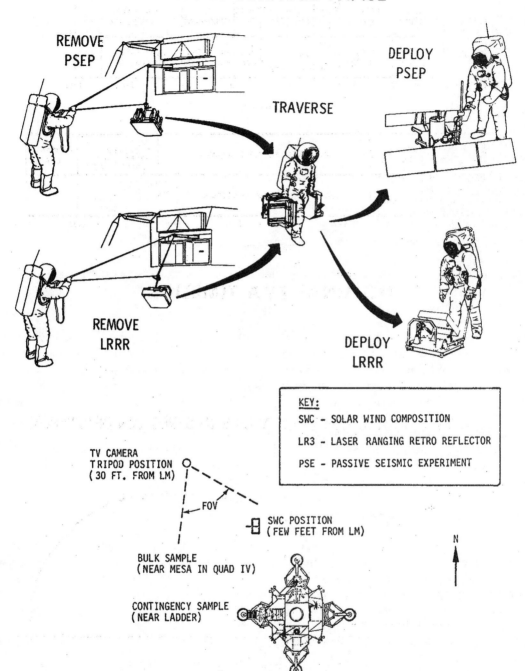

REMOVE
PSEP

TRAVERSE

DEPLOY
PSEP

REMOVE
LRRR

DEPLOY
LRRR

KEY:

SWC - SOLAR WIND COMPOSITION

LR3 - LASER RANGING RETRO REFLECTOR

PSE - PASSIVE SEISMIC EXPERIMENT

TV CAMERA
TRIPOD POSITION
(30 FT. FROM LM)

FOV

SWC POSITION
(FEW FEET FROM LM)

N

BULK SAMPLE
(NEAR MESA IN QUAD IV)

CONTINGENCY SAMPLE
(NEAR LADDER)

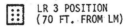

DOCUMENTED SAMPLE
(WITHIN 100 FT. FROM LM)

LR 3 POSITION
(70 FT. FROM LM)

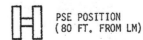

PSE POSITION
(80 FT. FROM LM)

Lunar Sample Collection

Equipment for collecting and stowing lunar surface samples is housed in the modularized equipment stowage assembly (MESA) on the LM descent stage. The commander will unstow the equipment after adjusting to the lunar surface environment.

Items stowed in the MESA are as follows:

* Black and white TV camera.

* Large scoop for collecting bulk and documented samples of loose lunar surface material.

* Extension handle that fits the large scoop, core tubes and hammer.

* Tongs for collecting samples of rock and for picking up dropped tools.

* Gnomon for vertical reference, color and dimension scale for lunar surface photography.

* Hammer for driving core tubes, chipping rock and for trenching (with extension handle attached).

* 35mm stereo camera.

* Two sample return containers (SRC) for returning up to 130 pounds of bulk and documented lunar samples. Items such as large and small sample bags, core tubes, gas analysis and lunar environment sample containers are stowed in the SRCs. Both containers are sealed after samples have been collected, documented and stowed, and the crew will hoist them into the ascent stage by means of an equipment conveyor for transfer into the command module and subsequent return to Earth for analysis in the Lunar Receiving Laboratory.

Additionally, a contingency lunar sample return container is stowed in the LM cabin for use by the commander during the early phases of his EVA. The device is a bag attached to an extending handle in which the commander will scoop up about one litre of lunar material. He then will jettison the handle and stow the contingency sample in his pressure suit pocket.

LM Ascent, Lunar Orbit Rendezvous

Following the 22-hour lunar stay time during which the commander and lunar module pilot will deploy the Early Apollo Scientific Experiments Package (EASEP), the Solar Wind Composition (SWC) experiment, and gather lunar soil samples, the LM ascent stage will lift off the lunar surface to begin the rendezvous sequence with the orbiting CSM. Ignition of the LM ascent engine will be at 124:23:21 for a 7 min 14 sec burn with a total velocity of 6,055 fps. Powered ascent is in two phases: vertical ascent for terrain clearance and the orbital insertion phase. Pitchover along the desired launch azimuth begins as the vertical ascent rate reached 50 fps about 10 seconds after liftoff at about 250 feet in altitude. Insertion into a 9 x 45-nm lunar orbit will take place about 166 nm west of the landing site.

Following LM insertion into lunar orbit, the LM crew will compute onboard the four major maneuvers for rendezvous with the CSM which is about 255 nm ahead of the LM at this point. All maneuvers in the sequence will be made with the LM RCS thrusters. The premission rendezvous sequence maneuvers, times and velocities which likely will differ slightly in real time, are as follows:

LM ASCENT

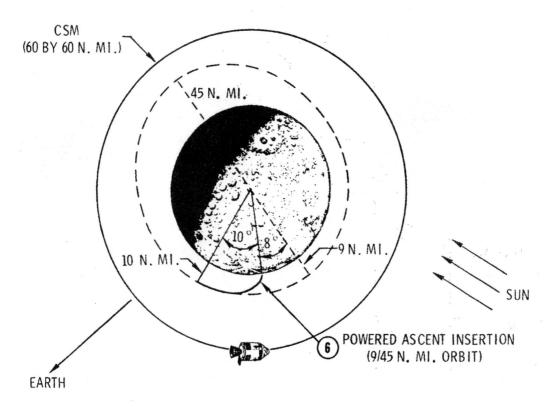

CSM
(60 BY 60 N. MI.)

45 N. MI.

10°   8

9 N. MI.

10 N. MI.

SUN

⑥ POWERED ASCENT INSERTION
(9/45 N. MI. ORBIT)

EARTH

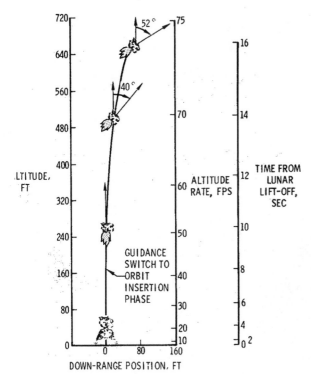

DOWN-RANGE POSITION, FT

VERTICAL RISE PHASE

Concentric sequence initiate (CSI): At first LM apolune after insertion 125:21:20 GET, 49 fps posigrade, following some 20 minutes of LM rendezvous radar tracking and CSM sextant/ VHF ranging navigation. CSI will be targeted to place the LM in an orbit 15 nm below the CSM at the time of the later constant delta height (CDH) maneuver. The CSI burn may also initiate corrections for any out-of-plane dispersions resulting from insertion azimuth errors. Resulting LM orbit after CSI will be 45.5 x 44.2 nm and will have a catchup rate to the CSM of .072 degrees per minute.

Another plane correction is possible about 29 minutes after CSI at the nodal crossing of the CSM and LM orbits to place both vehicles at a common node at the time of the CDH maneuver at 126:19:40 GET.

Terminal phase initiate (TPI): This maneuver occurs at 126:58:26 and adds 24.6 fps along the line of sight toward the CSM when the elevation angle to the CSM reaches 26.6 degrees. The LM orbit becomes 61.2 x 43.2 nm and the catchup rate to the CSM decreases to .032 degrees per second, or a closing rate of 131 fps.

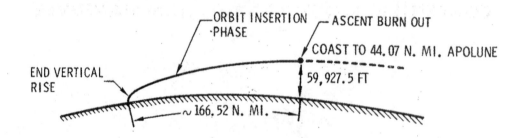

TOTAL ASCENT:
BURN TIME = 7:14.65 MIN:SEC
ΔV REQUIRED = 6,055.39 FPS
PROPELLANT REQUIRED = 4,989.86 LB

INSERTION ORBIT PARAMETERS

h_p = 55,905.4 FT

h_a = 44.07 N. MI.

η = 17.59°
γ = .324

ONBOARD DISPLAYS AT INSERTION

V = 5,535.9 FPS
h =   32.2 FPS
h = 60,129.5 FT

ORBIT INSERTION PHASE

Two midcourse correction maneuvers will be made if needed, followed by four braking maneuvers at: 127:39:43 GET, 11.5 fps; 127:40:56, 9.8 fps; 127:42:35 GET, 4.8 fps; and at 127:43:54 GET, 4.7 fps. Docking nominally will take place at 128 hrs GET to end three and one-half hours of the rendezvous sequence.

Transearth Injection (TEI)

The LM ascent stage will be jettisoned about four hours after hard docking and the CSM will make a 1 fps retrograde separation maneuver.

The nominal transearth injection burn will be at 135:24 GET following 59.5 hours in lunar orbit. TEI will take place on the lunar farside and will be a 3,293 fps posigrade SPS burn of 2 min 29 sec duration and will produce an entry velocity of 36,194 fps after a 59.6 hr transearth flight time.

An optional TEI plan for five revolutions later would allow a crew rest period before making the maneuver. TEI ignition under the optional plan would take place at 145:23:45 GET with a 3,698 fps posigrade SPS burn producing an entry velocity of 36,296 fps and a transearth flight time of 51.8 hrs.

LUNAR MODULE
CONCENTRIC SEQUENCE INITIATION MANEUVER

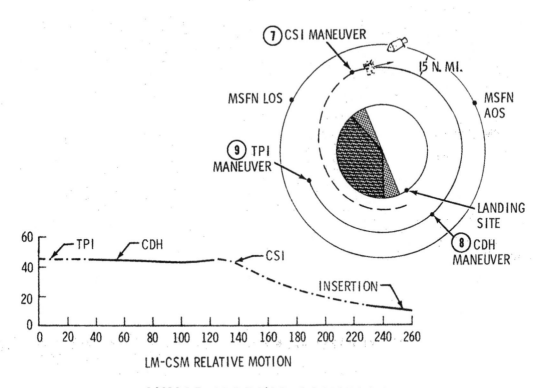

LUNAR MODULE CONSTANT
DIFFERENTIAL HEIGHT AND TERMINAL PHASE MANEUVERS

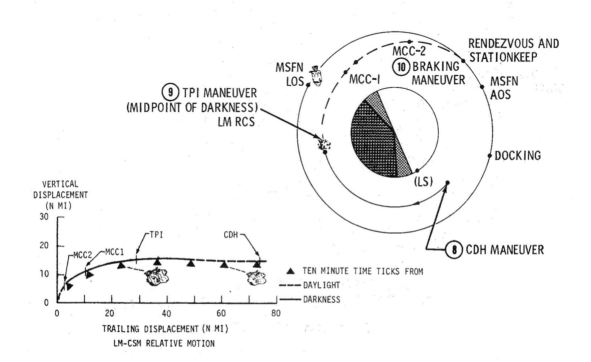

TRANSEARTH INJECTION

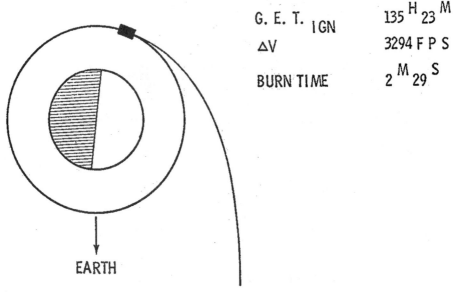

G. E. T. $_{IGN}$ 135^H 23^M

ΔV 3294 F P S

BURN TIME   2^M 29^S

EARTH

<u>Transearth coast</u>

Three corridor-control transearth Midcourse correction burns will be made if needed: MCC-5 at TEI +15 hrs, MCC-6 at entry interface (EI=400,000 feet) -15 hrs and MCC-7 at EI -3 hrs.

<u>Entry, Landing</u>

Apollo 11 will encounter the Earth's atmosphere (400,000 feet) at 195:05:04 GET at a velocity of 36,194 fps and will land some 1,285 nm downrange from the entry-interface point using the spacecraft's lifting characteristics to reach the landing point. Touchdown will be at 195:19:05 at 10.6 degrees north latitude by 172.4 west longitude.

EARTH ENTRY

ENTRY RANGE CAPABILITY - 1200 TO 2500 N. Mi.

NOMINAL ENTRY RANGE - 1285 N. Mi.

SHORT RANGE SELECTED FOR NOMINAL MISSION BECAUSE:
        RANGE FROM ENTRY TO LANDING CAN BE SAME FOR PRIMARY AND BACKUP CONTROL
        MODES

        PRIMARY MODE EASIER TO MONITOR WITH SHORT RANGE

WEATHER AVOIDANCE, WITHIN ONE DAY PRIOR TO ENTRY, IS ACHIEVED USING ENTRY RANGING CAPABILITY TO 2500 N. Mi.

UP TO ONE DAY PRIOR TO ENTRY USE PROPULSION SYSTEM
TO CHANGE LANDING POINT

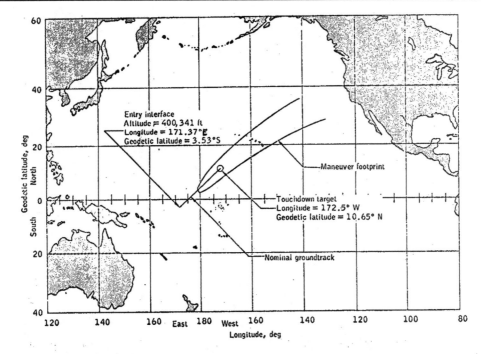

MANEUVER FOOTPRINT AND NOMINAL GROUNDTRACK

GEODETIC ALTITUDE VERSUS RANGE TO GO

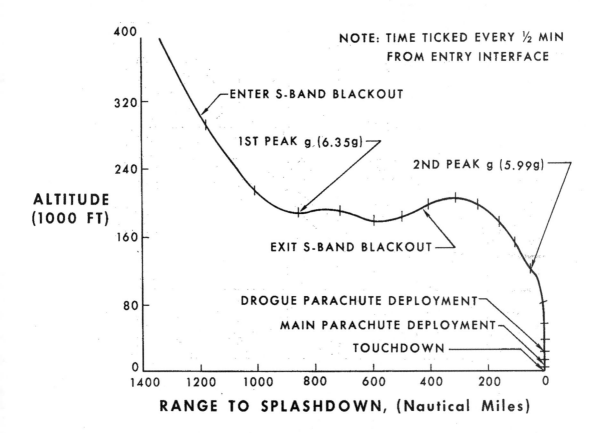

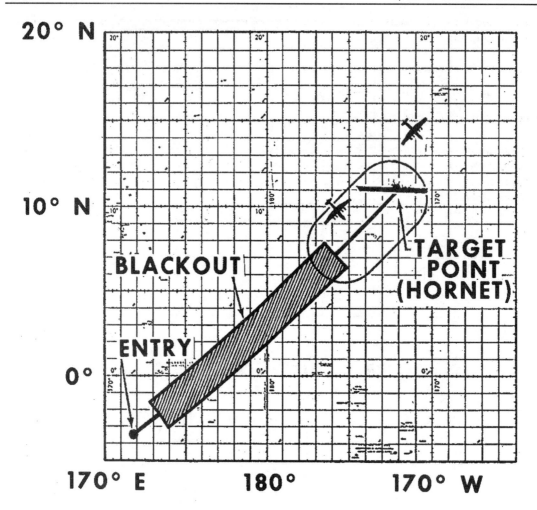

20° N

10° N

0°

170° E   **180°**   **170° W**

BLACKOUT

TARGET POINT (HORNET)

ENTRY

PRIMARY LANDING AREA

RECOVERY OPERATIONS, QUARANTINE

The prime recovery line for Apollo 11 is the mid-Pacific along the 175th west meridian of longitude above 15 degrees north latitude, and jogging to 165 degrees west longitude below the equator. The aircraft carrier USS Hornet, Apollo 11 prime recovery ship, will be stationed near the end-of-mission aiming point prior to entry.

Splashdown for a full-duration lunar landing mission launched on time July 16 will be at 10.6 degrees north by 172.5 degrees west at a ground elapsed time of 195 hrs 15 min.

The latitude of splashdown depends upon the time of the transearth injection burn and the declination of the Moon at the time of the burn. A spacecraft returning from a lunar mission will enter the Earth's atmosphere and splash down at a point on the Earth's farside directly opposite the Moon. This point, called the antipode, is a projection of a line from the center of the Moon through the center of the Earth to the surface opposite the Moon. The mid-Pacific recovery line rotates through the antipode once each 24 hours, and the transearth injection burn will be targeted for splashdown along the primary recovery line.

Other planned recovery lines for lunar missions are the East Pacific line extending roughly parallel to the coastlines of North and South America; the Atlantic Ocean line running along the 30th west meridian in the northern hemisphere and along the 25th west meridian in the southern hemisphere, and the Indian Ocean along the 65th east meridian.

Secondary landing areas for a possible Earth orbital alternate mission are in three zones — one in the Pacific and two in the Atlantic.

Launch abort landing areas extend downrange 3,200 nautical miles from Kennedy Space Center, fanwise 50 nm above and below the limits of the variable launch azimuth (72-106 degrees). Ships on station in the launch abort area will be the destroyer USS New, the insertion tracking ship USNS Vanguard and the minesweeper-countermeasures ship USS Ozark.

In addition to the primary recovery ship located on the mid-Pacific recovery line and surface vessels on the Atlantic Ocean recovery line and in the launch abort area, 13 HC-130 aircraft will be on standby at seven staging bases around the Earth: Guam; Hawaii; Bermuda; Lajes, Azores; Ascension Island; Mauritius and the Panama Canal Zone.

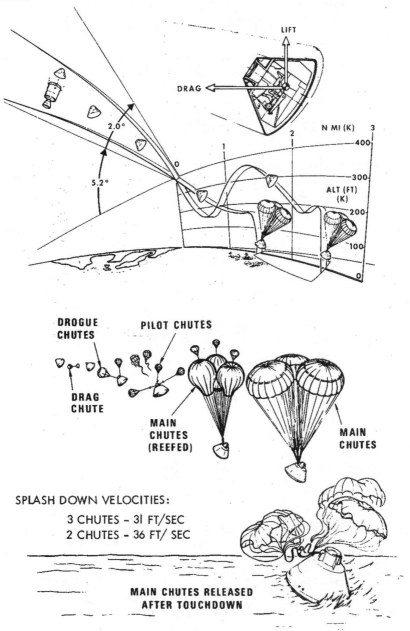

EARTH RE-ENTRY AND LANDING

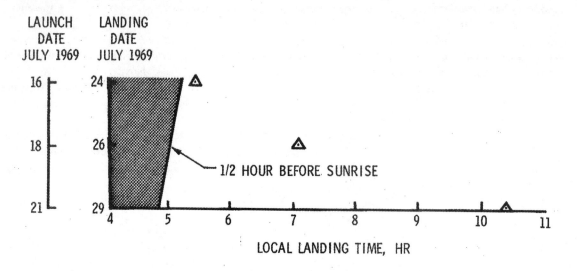

LOCAL LANDING TIME, HR

LOCAL LANDING TIMES

Apollo 11 recovery operations will be directed from the Recovery Operations Control Room in the Mission Control Center and will be supported by the Atlantic Recovery Control Center, Norfolk, Va., and the Pacific Recovery Control Center, Kunia, Hawaii.

After splashdown, the Apollo 11 crew will don biological isolation garments passed to them through the spacecraft hatch by a recovery swimmer. The crew will be carried by helicopter to the Hornet where they will enter a Mobile Quarantine Facility (MQF) about 90 minutes after landing. The MQF, with crew aboard, will be offloaded at Ford Island, Hawaii and loaded on a C-141 aircraft for the flight to Ellington AFB, Texas, and thence trucked to the Lunar Receiving Laboratory (LRL).

The crew will arrive at the LRL on July 27 following a nominal lunar landing mission and will go into the LRL Crew Reception area for a total of 21 days quarantine starting from the time they lifted off the lunar surface. The command module will arrive at the LRL two or three days later to undergo a similar quarantine. Lunar material samples will undergo a concurrent analysis in the LRL Sample Operations area during the quarantine period.

<u>Lunar Receiving Laboratory</u>

The Manned Spacecraft Center Lunar Receiving Laboratory has as its main function the quarantine and testing of lunar samples, spacecraft and flight crews for possible harmful organisms brought back from the lunar surface.

Detailed analysis of returned lunar samples will be done in two phases — time-critical investigations within the quarantine period and post-quarantine scientific studies of lunar samples repackaged and distributed to participating scientists
There are 36 scientists and scientific groups selected in open world-wide competition on the scientific merits of their proposed experiments. They represent some 20 institutions in Australia, Belgium, Canada, Finland, Federal Republic of Germany, Japan, Switzerland and the United Kingdom. Major fields of investigation will be mineralogy and petrology, chemical and isotope analysis, physical properties, and biochemical and organic analysis.

The crew reception area serves as quarters for the flight crew and attendant technicians for the quarantine period in which the pilots will be debriefed and examined. The other crew reception area occupants are physicians, medical technicians, housekeepers and cooks. The CRA is also a contingency quarantine area for sample operations area people exposed to spills or vacuum system breaks.

Both the crew reception area and the sample operations area are contained within biological barrier systems that protect lunar materials from Earth contamination as well as protect the outside world from possible contamination by lunar materials.

Analysis of lunar samples will be done in the sample operations area, and will include vacuum, magnetics, gas analysis, biological test, radiation counting and physical-chemical test laboratories.

Lunar sample return containers, or "rock boxes", will first be brought to the vacuum laboratory and opened in the ultra-clean vacuum system. After preliminary examination, the samples will be repackaged for transfer, still under vacuum, to the gas analysis, biological preparation, physical-chemical test and radiation counting laboratories.

The gas analysis lab will measure amounts and types of gases produced by lunar samples, and geochemists in the physical-chemical test lab will test the samples for their reactions to atmospheric gases and water vapor. Additionally, the physical-chemical test lab will make detailed studies of the mineralogic, petrologic, geochemical and physical properties of the samples.

Other portions of lunar samples will travel through the LRL vacuum system to the biological test lab where they will undergo tests to determine if there is life in the material that may replicate. These tests will involve introduction of lunar samples into small germ-free animals and plants. The biological test laboratory is made up of several smaller labs — bioprep,

BIOLOGICAL ISOLATION GARMENT

bioanalysis, germ-free, histology, normal animals (amphibia and invertebrates), incubation, anaerobic and tissue culture, crew microbiology and plants.

Some 50 feet below the LRL ground floor, the radiation counting lab will conduct low-background radioactive assay of lunar samples using gamma ray spectrometry techniques.

(See Contamination Control Program section for more details on LRL, BIGs, and the Mobile Quarantine Facility.)

SCHEDULE FOR TRANSPORT OF SAMPLES, SPACECRAFT, CREW

Samples

Two helicopters will carry lunar samples from the recovery ship to Johnston Island where they will be put aboard a C-141 and flown directly to Houston and the Lunar Receiving Laboratory (LRL). The samples should arrive at Ellington Air Force Base at about 27 hours after recovery and received in the LRL at about 9 or 10 a.m. CDT, July 25.

Spacecraft

The spacecraft is scheduled to be brought aboard the recovery ship about two hours after recovery. About 55 hours after recovery the ship is expected to arrive in Hawaii. The spacecraft will be deactivated in Hawaii (Ford Island) between 55 and 127 hours after recovery. At 130 hours it is scheduled to be loaded on a C-133 for return to Ellington AFB. Estimated time of arrival at the LRL is on July 29, 140 hours after recovery.

Crew

The flight crew is expected to enter the Mobile Quarantine Facility (MQF) on the recovery ship about 90 minutes after splashdown. The ship is expected to arrive in Hawaii at recovery plus 55 hours and the Mobile Quarantine Facility will be transferred to a C-141 aircraft at recovery Plus 57 hours. The aircraft will land at Ellington AFB at recovery plus 65 hours and the MQF will arrive at the LRL about two hours later (July 27).

LUNAR RECEIVING LABORATORY PROCEDURES TIMELINE (TENTATIVE)

Sample Operations Area (SOAO)

Arrival LRL	Event	Location
Arrival	Sample containers arrive crew reception area, outer covering checked, tapes and films removed	Crew reception area
Arrival	Container #1 introduced into system. Containers weighed Transfer contingency sample to F-25a chamber for examination after containers #1 and #2 Containers sterilized, dried in atmospheric decontamination and passed into glove chamber F201 Residual gas analyzed (from containers)	Vacuum chamber lab
Plus 5 hours	Open containers, Weigh, preliminary exam of samples and first visual inspection by preliminary evaluation team	
Plus 8 hours	Remove samples to Radiation Counting, Gas Analysis Lab & Mineralogy & Petrology Lab	Vacuum chamber lab RCL Basement Min-Pet 1st floor
Plus 13 hours	Preliminary information Radiation counting. Transfer container #1 out of chamber	Vacuum chamber lab

Arrival LRL	Event	Location
	Initial detailed exam by Preliminary Evaluation Team Members	Vacuum chamber lab
	Sterile sample to Bio prep (100 gms) (24 to 48 hr preparation for analysis)	Bio Test area - 1st floor
	Monopole experiment	Vacuum chamber lab
Plus 13 hours	Transfer samples to Phys-Chem Lab	Phys-Chem - 1st floor
	Detailed photography of samples and microscopic work	Vacuum chamber lab
Plus 24 hours	All samples canned and remain in chamber	
Plus 1-2 days	Preparations of samples in bioprep lab for distribution to bio test labs. (Bacteriology, Virology, Germ-free mice) through TEI plus 21 days	Bio test labs - 1st floor
Plus 4-5 days	Early release of phys-chem analysis	Phys-Chem labs - 1st floor
Plus about 7 - 15 days	Detailed bio analysis & further phys-chem analysis	Bio test & min-pet 1st floor
Plus 15 days	Conventional samples transferred to bio test area (24-48 hours preparation for analysis)	1st floor
Plus 17 days	Bio test begins on additional bacteriological, virological, microbiological invertebrates, (fish, shrimp, oysters), birds, mice, lower invertebrates (house fly, moth, german cockroach, etc), plants (about 20) (through approximately arrival Plus 30 days)	1st floor
Plus 30 days	Bio test info released on pre-liminary findings	1st floor
	Samples go to thin section lab (first time outside barrier) for preparation and shipment to principal investigators	1st floor

APOLLO 11 GO/NO-GO DECISION POINTS

Like Apollo 8 and 10, Apollo 11 will be flown on a step-by-step commit point or go/no-go basis in which the decisions will be made prior to each maneuver whether to continue the mission or to switch to one of the possible alternate missions. The go/no-go decisions will be made by the flight control teams in Mission Control Center jointly with the flight crew.

Go/no-go decisions will be made prior to the following events:

*   Launch phase go/no-go at 10 min GET for orbit insertion

*   Translunar injection

*   Transposition, docking and LM extraction

*   Each translunar midcourse correction burn

*   Lunar orbit insertion burns Nos. 1 and 2

*   CSM-LM undocking and separation

*   LM descent orbit insertion

*   LM powered descent initiation

*   LM landing

*   Periodic go/no-gos during lunar stay

*   Lunar surface extravehicular activity

*   LM ascent and rendezvous (A no-go would delay ascent one revolution)

*   Transearth injection burn (no-go would delay TEI one or more revolutions to allow maneuver preparations to be completed)

*   Each transearth midcourse correction burn.

APOLLO 11 ALTERNATE MISSIONS

Six Apollo 11 alternate missions, each aimed toward meeting the maximum number of mission objectives and gaining maximum Apollo systems experience, have been evolved for realtime choice by the mission director. The alternate missions are summarized as follows:

Alternate 1 - S-IVB fails prior to Earth orbit insertion: CSM only contingency orbit insertion (COI) with service propulsion system. The mission in Earth orbit would follow the lunar mission timeline as closely as possible and would include SPS burns similar in duration to LOI and TEI, while at the same time retaining an RCS deorbit capability. Landing would be targeted as closely as possible to the original aiming point.

Alternate 2 - S-IVB fails to restart for TLI: CSM would dock with and extract the LM as soon as possible and perform an Earth orbit mission, including docked DPS burns and possibly CSM-active rendezvous along the lunar mission timeline, with landing at the original aiming point. Failure to extract the LM would result in an Alternate 1 type mission.

Alternate 3 - No-go for nominal TLI because of orbital conditions or insufficient S-IVB propellants: TLI retargeted for lunar mission if possible; if not possible, Alternate 2 would be followed. The S-IVB would be restarted for a high-ellipse injection provided an apogee greater than 35,000 nm could be achieved. If propellants available in the S-IVB were too low to reach the 35,000 nm apogee, the TLI burn would be targeted out of plane and an Earth orbit mission along the lunar mission timeline would be flown.

Depending upon the quantity of S-IVB propellant available for a TLI-type burn that would produce an apogee greater than 35,000 nm, Alternate 3 is broken down into four subalternates:

Alternate 3A - Propellant insufficient to reach 35,000 nm

Alternate 3B -   Propellant sufficient to reach apogee between 35,000 and 65,000 nm

Alternate 3C -   Propellant sufficient to reach apogee between 65,000 and 200,000 nm

Alternate 3D -   Propellant sufficient to reach apogee of 200,000 nm or greater; this alternate would be a near-nominal TLI burn and midcourse correction burn No. I would be targeted to adjust to a free-return trajectory.

Alternate 4 - Non-nominal or early shutdown TLI turn: Real-time decision would be made on whether to attempt a lunar mission or an Earth orbit mission, depending upon when TLI cutoff occurs. A lunar mission would be possible if cutoff took place during the last 40 to 45 seconds of the TLI burn. Any alternate mission chosen would include adjusting the trajectory to fit one of the above listed alternates and touchdown at the nominal mid-Pacific target point.

Alternate 5 - Failure of LM to eject after transposition and docking: CSM would continue alone for a circumlunar or lunar orbit mission, depending upon spacecraft systems status.

Alternate 6 - LM systems failure in lunar orbit: Mission would be modified in real time to gain the maximum of LM systems experience within limits of crew safety and time. If the LM descent propulsion system operated normally, the LM would be retained for DPS backup transearth injection; if the DPS were no-go, the entire LM would be jettisoned prior to TEI.

ABORT MODES

The Apollo 11 mission can be aborted at any time during the launch phase or terminated during later phases after a successful insertion into Earth orbit.

Abort modes can be summarized as follows:

Launch phase —

Mode I — Launch escape system (LES) tower propels command module away from launch vehicle. This mode is in effect from about T-45 minutes when LES is armed until LES tower jettison at 3:07 GET and command module landing point can range from the Launch Complex 39A area to 400 nm downrange.

Mode II — Begins when LES tower is jettisoned and runs until the SPS can be used to insert the CSM into a safe Earth orbit (9:22 GET) or until landing points approach the African coast. Mode II requires manual separation, entry orientation and full-lift entry with landing between 350 and 3,200 nm downrange.

Mode III — Begins when full-lift landing point reached 3,200 nm (3,560 sm, 5,931 km) and extends through Earth orbital insertion. The CSM would separate from the launch vehicle, and if necessary, an SPS retrograde burn would be made, and the command module would be flown half-lift to entry and landing at approximately 3,350 nm (3,852 sm, 6,197 km) downrange.

Mode IV and Apogee Kick — Begins after the point the SPS could be used to insert the CSM into an Earth parking orbit -from about 9:22 GET. The SPS burn into orbit would be made two minutes after separation from the S-IVB and the mission would continue as an Earth orbit alternate. Mode IV is preferred over Mode III. A variation of Mode IV is the apogee kick in which the SPS would be ignited at first apogee to raise perigee for a safe orbit.

Deep Space Aborts

Translunar Injection Phase —

Aborts during the translunar injection phase are only a remote possibility, but if an abort became necessary during the TLI maneuver, an SPS retrograde burn could be made to produce spacecraft entry. This mode of abort would be used only in the event of an extreme emergency that affected crew safety. The spacecraft landing point would vary with launch azimuth and length of the TLI burn. Another TLI abort situation would be used if a malfunction cropped up after injection. A retrograde SPS burn at about 90 minutes after TLI shutoff would allow targeting to land on the Atlantic Ocean recovery line.

Translunar Coast phase —

Aborts arising during the three-day translunar coast phase would be similar in nature to the 90-minute TLI abort. Aborts from deep space bring into the play the Moon's antipode (line projected from Moon's center through Earth's Center to the surface opposite the Moon) and the effect of the Earth's rotation upon the geographical location of the antipode. Abort times would be selected for landing when the 165 degree west longitude line crosses the antipode. The mid-Pacific recovery line crosses the antipode once each 24 hours, and if a time-critical situation forces an abort earlier than the selected fixed abort times, landings would be targeted for the Atlantic Ocean, West Pacific or Indian Ocean recovery lines in that order of preference. When the spacecraft enters the Moon's sphere of influence, a circumlunar abort becomes faster than an attempt to return directly to Earth.

Lunar Orbit Insertion phase —

Early SPS shutdowns during the lunar orbit insertion burn (LOI) are covered by three modes in the Apollo 11 mission. All three modes would result in the CM landing at the Earth latitude of the Moon antipode at the time the abort was performed.

Mode I would be a LM DPS posigrade burn into an Earth return trajectory about two hours (at next pericynthion) after an LOI shutdown during the first two minutes of the LOI burn.

Mode II, for SPS shutdown between two and three minutes after ignition, would use the LM DPS engine to adjust the orbit to a safe, non-lunar impact trajectory followed by a second DPS posigrade burn at next pericynthion targeted for the mid-Pacific recovery line.

Mode III, from three minutes after LOI ignition until normal cutoff, would allow the spacecraft to coast through one or two lunar orbits before doing a DPS posigrade burn at pericynthion targeted for the mid-Pacific recovery line.

Lunar Orbit Phase —

If during lunar parking orbit it became necessary to abort, the transearth injection (TEI) burn would be made early and would target spacecraft landing to the mid-Pacific recovery line.

Transearth Injection phase —

Early shutdown of the TEI burn between ignition and two minutes would cause a Mode III abort and a SPS posigrade TEI burn would be made at a later pericynthion. Cutoffs after two minutes TEI burn time would call for a Mode I abort— restart of SPS as soon as possible for Earth return trajectory. Both modes produce mid-Pacific recovery line landings near the latitude of the antipode at the time of the TEI burn.

Transearth Coast phase —

Adjustments of the landing point are possible during the transearth coast through burns with the SPS or the service module RCS thrusters, but in general, these are covered in the discussion of transearth midcourse corrections. No abort burns will be made later than 24 hours prior to entry to avoid effects upon CM entry velocity and flight path angle.

APOLLO 11 ONBOARD TELEVISION

Two television cameras will be carried aboard Apollo 11. A color camera of the type used on Apollo 10 will be stowed for use aboard the command module, and the black-and-white Apollo lunar television camera will be stowed in the LM descent stage for televising back to Earth a real-time record of man's first step onto the Moon.

The lunar television camera weighs 7.25 pounds and draws 6.5 watts of 24-32 volts DC power. Scan rate is 10 frames-per-second at 320 lines-per-frame. The camera body is 10.6 inches long, 6.5 inches wide and 3.4 inches deep. The bayonet lens mount permits lens changes by a crewman in a pressurized suit. Two lenses, a wideangle lens for closeups and large areas, and a lunar day lens for viewing lunar surface features and activities in the near field of view with sunlight illumination, will be provided for the lunar TV camera.

The black-and-white lunar television camera is stowed in the MESA (Modular Equipment Stowage Assembly) in the LM descent stage and will be powered up before Armstrong starts down the LM ladder. When he pulls the lanyard to deploy the MESA, the TV camera will also swing down on the MESA to the left of the ladder (as viewed from LM front) and relay a TV picture of his initial steps on the Moon. Armstrong later will mount the TV camera on a tripod some distance away from the LM after Aldrin has descended to the surface. The camera will be left untended to cover the crew's activities during the remainder of the EVA.

The Apollo lunar television camera is built by Westinghouse Electric Corp., Aerospace Division, Baltimore, Md.

The color TV camera is a 12-pound Westinghouse camera with a zoom lens for wideangle or close-up use, and has a three inch monitor which can be mounted on the camera or in the command module. The color camera outputs a standard 525-line, 30 frame-per-second signal in color by use of a rotating color wheel. The black-and-white signal from the spacecraft will be converted to color at the Mission Control Center.

The following is a preliminary plan for TV passes based upon a 9:32 a.m. EDT, July 16 launch.

TENTATIVE APOLLO 11 TV TIMES

Date	Times of Planned TV (EDT)	GET	Prime Site	Event
July 17	7:32 - 7:47 P.m.	34:00-34:15	Goldstone	Translunar Coast
July 18	7:32 - 7:47 p.m.	58:00-58:15	Goldstone	Translunar Coast
July 19	4:02 - 4:17 P.m.	78:30-78:45	Goldstone	Lunar Orbit (general surface shots)
July 20	1:52 - 2:22 p.m.	100:20-100:50	Madrid	CM/LM Formation Flying
July 21	1:57 - 2:07 a.m.	112:25-112:35	Goldstone	Landing Site Tracking
July 21	2:12 - 4:52 a.m.	112:40-115:20	*Parkes	Black and White Lunar Surface
July 22	9:02 - 9:17 P.m.	155:30-155:45	Goldstone	Transearth Coast
July 23	7:02 - 7:17 P.m.	177:30-177:45	Goldstone	Transearth Coast

Honeysuckle will tape the Parkes pass and ship tape to MSC.

APOLLO 11 PHOTOGRAPHIC TASKS

Still and motion pictures will be made of most spacecraft maneuvers as well as of the lunar surface and of crew activities in the Apollo 11 cabin. During lunar surface activities after lunar module touchdown and the two hour 40 minute EVA, emphasis will be on photographic documentation of crew mobility, lunar surface features and lunar material sample collection.

Camera equipment carried on Apollo 11 consists of one 70mm Hasselblad electric camera stowed aboard the command module, two Hasselblad 70mm lunar surface superwide angle cameras stowed aboard the LM and a 35mm stereo close-up camera in the LM MESA.

The 2.3 pound Hasselblad superwide angle camera in the LM is fitted with a 38mm f/4-5 Zeiss Biogon lens with a focusing range from 12 inches to infinity. Shutter speeds range from time exposure and one second to 1/500 second. The angular field of view with the 38mm lens is 71 degrees vertical and horizontal on the square-format film frame.

The command module Hasselblad electric camera is normally fitted with an 80mm f/2.8 Zeiss Planar lens, but bayonet-mount 60mm and 250mm lens may be substituted for special tasks. The 80mm lens has a focusing range from three feet to infinity and has a field of view of 38 degrees vertical and horizontal.

Stowed with the Hasselblads are such associated items as a spotmeter, ringsight, polarizing filter, and film magazines. Both versions of the Hasselblad accept the same type film magazine.

For motion pictures, two Maurer 16mm data acquisition cameras (one in the CSM, one in the LM) with variable frame speed (1, 6, 12 and 24 frames per second) will be used. The cameras each weigh 2.8 pounds with a 130-foot film magazine attached. The command module 16mm camera will have lenses of 5, 18 and 75mm focal length available, while the LM camera will be fitted with the 18mm wideangle lens. Motion picture camera accessories include a right-angle mirror, a power cable and a command module boresight window bracket.

During the lunar surface extravehicular activity, the commander will be filmed by the LM pilot with the LM 16mm camera at normal or near-normal frame rates (24 and 12 fps), but when he leaves the LM to join the commander, he will switch to a one frame-per-second rate. The camera will be mounted inside the LM looking through the right-hand window. The 18mm lens has a horizontal field of view of 32 degrees and a vertical field of view of 23 degrees. At one fps, a 130-foot 16mm magazine will run out in 87 minutes in real time; projected at the standard 24 fps, the film would compress the 87 minutes to 3.6 minutes.

Armstrong and Aldrin will use the Hasselblad lunar surface camera extensively during their surface EVA to document each of their major tasks. Additionally, they will make a 360-degree overlapping panorama sequence of still photos of the lunar horizon, photograph surface features in the immediate area, make close-ups of geological samples and the area from which they were collected and record on film the appearance and condition of the lunar module after landing.

Stowed in the MESA is a 35mm stereo close-up camera which shoots 24mm square color stereo pairs with an image scale of one-half actual size. The camera is fixed focus and is equipped with a stand-off hood to position the camera at the proper focus distance. A long handle permits an EVA crewman to position the camera without stooping for surface object photography. Detail as small as 40 microns can be recorded.

A battery-powered electronic flash provides illumination. Film capacity is a minimum of 100 stereo pairs.

The stereo close-up camera will permit the Apollo 11 landing crew to photograph significant surface structure phenomena which would remain intact only in the lunar environment, such as fine powdery deposits, cracks or holes and adhesion of particles.

Near the end of EVA, the film cassette will be removed and stowed in the commander's contingency sample container pocket and the camera body will be left on the lunar surface.

LUNAR DESCRIPTION

Terrain - Mountainous and crater-pitted, the former rising thousands of feet and the latter ranging from a few inches to 180 miles in diameter. The craters are thought to be formed by the impact of meteorites. The surface is covered with a layer of fine-grained material resembling silt or sand, as well as small rocks and boulders.

Environment - No air, no wind, and no moisture. The temperature ranges from 243 degrees in the two-week lunar day to 279 degrees below zero in the two-week lunar night. Gravity is one-sixth that of Earth. Micrometeoroids pelt the Moon (there is no atmosphere to burn them up). Radiation might present a problem during periods of unusual solar activity.

Dark Side - The dark or hidden side of the Moon no longer is a complete mystery. It was first photographed by a Russian craft and since then has been photographed many times, particularly by NASA's Lunar Orbiter spacecraft and Apollo 8.

Origin - There is still no agreement among scientists on the origin of the Moon. The three theories:
(1) the Moon once was part of Earth and split off into its own orbit,
(2) It evolved as a separate body at the same time as Earth, and
(3) it formed elsewhere in space and wandered until it was captured by Earth's gravitational field.

Physical Facts

Diameter	2,160 miles (about ¼ that of Earth)
Circumference	6,790 miles (about ¼ that of Earth)
Distance from Earth	238,857 miles (mean; 221,463 minimum to 252,710 maximum)
Surface temperature	+243°F (Sun at zenith) -279°F (night)
Surface gravity	1/6 that of Earth
Mass	1/100th that of Earth
Volume	1/50th that of Earth
Lunar day and night	14 Earth days each
Mean velocity in orbit	2,287 miles per hour
Escape velocity	1.48 miles per second
Month (period of rotation around Earth)	27 days, 7 hours, 43 minutes

APOLLO LUNAR LANDING SITES

Possible landing sites for the Apollo lunar module have been under study by NASA's Apollo Site Selection Board for more than two years. Thirty sites originally were considered. These have been narrowed down to three for the first lunar landing. (Site I currently not considered for first landing.)

Selection of the final sites was based on high resolution photographs by Lunar Orbiter spacecraft, plus close-up photos and surface data provided by the Surveyor spacecraft which soft-landed on the Moon.

The original sites are located on the visible side of the Moon within 45 degrees east and west of the Moon's center and 5 degrees north and south of its equator.

The final site choices were based on these factors:

* Smoothness (relatively few craters and boulders)

* Approach (no large hills, high cliffs, or deep craters that could cause incorrect altitude signals to the lunar module landing radar)

* Propellant requirements (selected sites require the least expenditure of spacecraft propellants)

* Recycle (selected sites allow effective launch preparation recycling if the Apollo Saturn V countdown is delayed)

* Free return (sites are within reach of the spacecraft launched on a free return translunar trajectory)

* Slope (there is little slope — less than 2 degrees in the approach path and landing area)

APOLLO LUNAR LANDING SITES

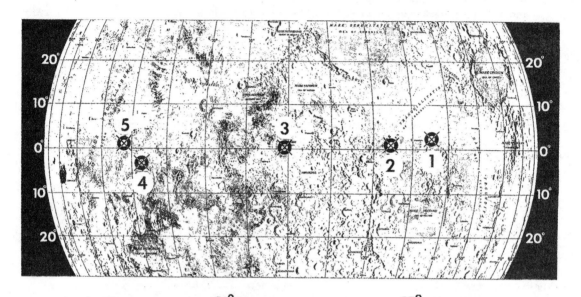

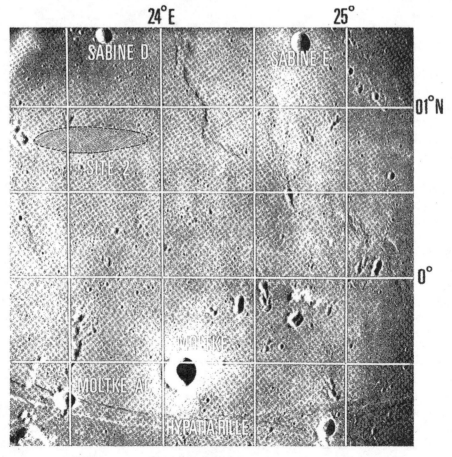

The Apollo 11 Landing Sites Are:

Site 2   latitude 0° 42' 50" North — longitude 23° 42' 28" East
          Site 2 is located on the east central part of the Moon in south
          western Mare Tranquillitatis. The site is approximately 62 miles
          (100 kilometers) east of the rim of Crater Sabine and approximately
          118 miles (190 kilometers) south west of the Crater Maskelyne.

Site 3   latitude 0° 21' 10" North — longitude 1° 171 57" West
          Site 3 is located near the center of the visible face of the Moon
          in the southwestern part of Sinus Medii. The site is approximately
          25 miles (40 kilometers) west of the center of the face and 21 miles
          (50 kilometers) southwest of the Crater Bruce.

Site 5   latitude 1° 40' 41" North — longitude 41° 53' 57" West
          Site 5 is located on the west central part of the visible face
          in southeastern Oceanus Procellarum. The site is approximately
          130 miles (210 kilometers) south west of the rim of Crater Kepler
          and 118 miles (190 kilometers) north northeast of the rim of
          Crater Flamsteed.

COMMAND AND SERVICE MODULE STRUCTURE, SYSTEMS

The Apollo spacecraft for the Apollo 11 mission is comprised of Command Module 107, Service Module 107, Lunar Module 5, a spacecraft-lunar module adapter (SLA) and a launch escape system. The SLA serves as a mating structure between the instrument unit atop the S-IVB stage of the Saturn V launch vehicle and as a housing for the lunar module.

Launch Escape System (LES) — Propels command module to safety in an aborted launch. It is made up of an open-frame tower structure, mounted to the command module by four frangible bolts, and three solid-propellant rocket motors: a 147,000 pound-thrust launch escape system motor, a 2,400 pound-thrust Ditch control motor, and a 31,500-pound-thrust tower jettison motor. Two canard vanes near the top deploy to turn the command module aerodynamically to an attitude with the heatshield forward. Attached to the base of the launch escape tower is a boost protective cover composed of resin impregnated fiberglass covered with cork, that protects the command module from aerodynamic heating during boost and rocket exhaust gases from the main and the jettison motors. The system is 33 feet tall, four feet in diameter at the base, and weighs 8,910 pounds.

Command Module (CM) Structure — The basic structure of the command module is a pressure vessel encased in heat shields, coneshaped 11 feet 5 inches high, base diameter of 12 feet 10 inches, and launch weight 12,250 pounds.

The command module consists of the forward compartment which contains two reaction control engines and components of the Earth landing system; the crew compartment or inner pressure vessel containing crew accomodations, controls and displays, and many of the spacecraft systems; and the aft compartment housing ten reaction control engines, propellant tankage, helium tanks, water tanks, and the CSM umbilical cable. The crew compartment contains 210 cubic feet of habitable volume.

Heat-shields around the three compartments are made of brazed stainless steel honeycomb with an outer layer of phenolic epoxy resin as an ablative material. Shield thickness, varying according to heat loads, ranges from 0.7 inch at the apex to 2.7 inches at the aft end.

The spacecraft inner structure is of sheet-aluminum honeycomb bonded sandwich ranging in thickness from 0.25 inch thick at forward access tunnel to 1.5 inches thick at base.

APOLLO SPACECRAFT

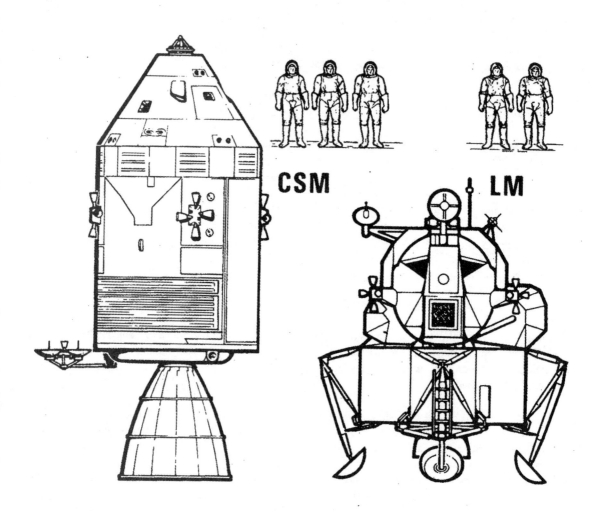

CSM   **LM**

CSM 107 and LM-5 are equipped with the probe-and-drogue docking hardware. The probe assembly is a powered folding coupling and impact attentuating device mounted on the CM tunnel that mates with a conical drogue mounted in the LM docking tunnel. After the 12 automatic docking latches are checked following a docking maneuver, both the probe and drogue assemblies are removed from the vehicle tunnels and stowed to allow free crew transfer between the CSM and LM.

Service Module (SM) Structure — The service module is a cylinder 12 feet 10 inches in diameter by 24 feet 7 inches high. For the Apollo 11 mission, it will weigh, 51,243 pounds at launch. Aluminum honeycomb panels one inch thick form the outer skin, and milled aluminum radial beams separate the interior into six sections around a central cylinder containing two helium spheres, four sections containing service propulsion system fuel-oxidizer tankage, another containing fuel cells, cryogenic oxygen and hydrogen, and one sector essentially empty.

Spacecraft LM Adapter (SLA) Structure — The spacecraft LM adapter is a truncated cone 28 feet long tapering from 260 inches diameter at the base to 154 inches at the forward end at the service module mating line. Aluminum honeycomb 1.75 inches thick is the stressed-skin structure for the spacecraft adapter. The SLA weighs 4,000 pounds.

CSM Systems

Guidance, Navigation and Control System (GNCS) — Measures and controls spacecraft position, attitude, and velocity, calculates trajectory, controls spacecraft propulsion system thrust vector, and displays abort data. The guidance system consists of three subsystems: inertial, made up of an inertial measurement unit and associated power and data components; computer which processes information to or from other components; and optics, including scanning telescope and sextant for celestial and/or landmark spacecraft navigation. CSM 107 and subsequent modules are equipped with a VHF ranging device as a backup to the LM rendezvous radar.

Stabilization and Control Systems (SCS) — Controls spacecraft rotation, translation, and thrust vector and provides displays for crew-initiated maneuvers; backs up the guidance system. It has three subsystems; attitude reference, attitude control, and thrust vector control.

Service Propulsion System (SPS) — Provides thrust for large spacecraft velocity changes through a gimbal-mounted 20,500-pound-thrust hypergolic engine using a nitrogen tetroxide oxidizer and a 50-50 mixture of unsymmetrical dimethyl hydrazine and hydrazine fuel. This system is in the service module. The system responds to automatic firing commands from the guidance and navigation system or to manual commands from the crew. The engine provides a constant thrust level. The stabilization and control system gimbals the engine to direct the thrust vector through the spacecraft center of gravity.

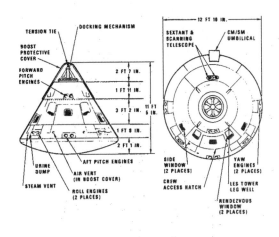

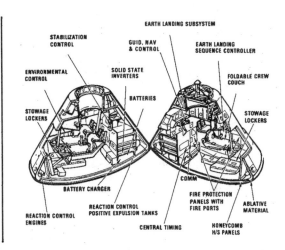

COMMAND MODULE

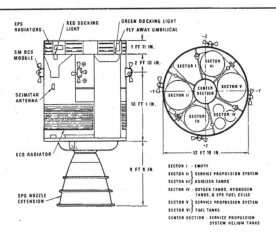

SERVICE MODULE

Telecommunications System — Provides voice, television, telemetry, and command data and tracking and ranging between the spacecraft and Earth, between the command module and the lunar module and between the spacecraft and the extravehicular astronaut. It also provides intercommunications between astronauts. The telecommunications system consists of pulse code modulated telemetry for relaying to Manned Space Flight Network stations data on spacecraft systems and crew condition, VHF/AM voice, and unified S-Band tracking transponder, air-to-ground voice communications, onboard television, and a VHF recovery beacon. Network stations can transmit to the spacecraft such items as updates to the Apollo guidance computer and central timing equipment, and real-time commands for certain onboard functions.

The high-gain steerable S-Band antenna consists of four, 31-inch-diameter parabolic dishes mounted on a folding boom at the aft end of the service module. Nested alongside the service propulsion system engine nozzle until deployment, the antenna swings out at right angles to the spacecraft longitudinal axis, with the boom pointing 52 degrees below the heads-up horizontal. Signals from the ground stations can be tracked either automatically or manually with the antenna's gimballing system. Normal S-Band voice and uplink/downlink communications will be handled by the omni and high-gain antennas.

Sequential System — Interfaces with other spacecraft systems and subsystems to initiate time critical functions during launch, docking maneuvers, sub-orbital aborts, and entry portions of a mission. The system also controls routine spacecraft sequencing such as service module separation and deployment of the Earth landing system.

Emergency Detection System (EDS) — Detects and displays to the crew launch vehicle emergency conditions, such as excessive pitch or roll rates or two engines out, and automatically or manually shuts down the booster and activates the launch escape system; functions until the spacecraft is in orbit.

Earth Landing System (ELS) — Includes the drogue and main parachute system as well as post-landing recovery aids. In a normal entry descent, the command module forward heat shield is jettisoned at 24,000 feet, permitting mortar deployment of two reefed 16.5-foot diameter drogue parachutes for orienting and decelerating the spacecraft. After disreef and drogue release, three mortar deployed pilot chutes pull out the three main 83.3 foot diameter parachutes with two-stage reefing to provide gradual inflation in three steps. Two main parachutes out of three can provide a safe landing.

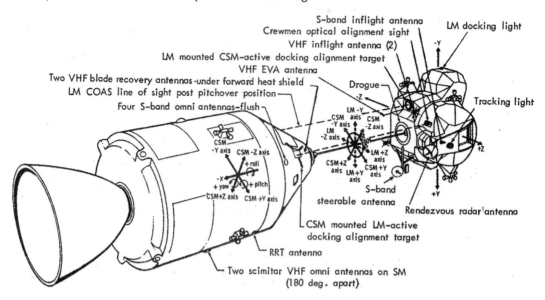

SPACECRAFT AXIS AND ANTENNA LOCATIONS

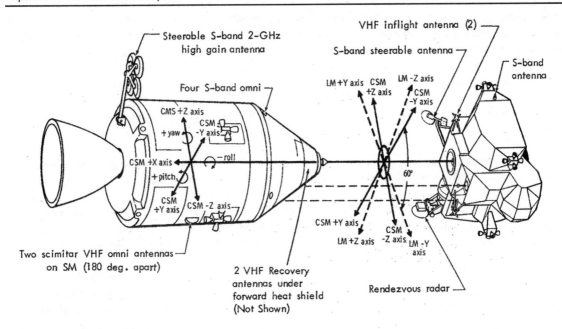

Steerable S-band 2-GHz high gain antenna

Four S-band omni

CMS +Z axis

CSM -Y axis

CSM +yaw

CSM +X axis — roll

CSM +pitch

CSM +Y axis   CSM -Z axis

Two scimitar VHF omni antennas on SM (180 deg. apart)

2 VHF Recovery antennas under forward heat shield (Not Shown)

VHF inflight antenna (2)

S-band steerable antenna

S-band antenna

LM +Y axis CSM +Z axis   LM -Z axis   CSM -Y axis

60°

CSM +Y axis   LM +Z axis   CSM -Z axis   LM -Y axis

Rendezvous radar

SPACECRAFT AXIS AND ANTENNA LOCATIONS

Reaction Control System (RCS) — The command module and the service module each has its own independent system. The SM RCS has four identical RCS "quads" mounted around the SM 90 degrees apart. Each quad has four 100 pound-thrust engines, two fuel and two oxidizer tanks and a helium pressurization sphere. The SM RCS provides redundant spacecraft attitude control through crosscoupling logic inputs from the stabilization and guidance systems. Small velocity change maneuvers can also be made with the SM RCS.

The CM RCS consists of two independent six-engine subsystems of six 93 poundthrust engines each. Both subsystems are activated just prior to CM separation from the SM: one is used for spacecraft attitude control during entry. The other serves in standby as a backup. Propellants for both CM and SM RCS are monomethyl hydrazine fuel and nitrogen tetroxide oxidizer with helium pressurization. These propellants are hypergolic, i.e. they burn spontaneously when combined without an igniter.

Electrical Power System (EPS) — Provides electrical energy sources, power generation and control, power conversion and conditioning, and power distribution to the spacecraft throughout the mission. The EPS also furnishes drinking water to the astronauts as a byproduct of the fuel cells. The primary source of electrical power is the fuel cells mounted in the SM. Each cell consists of a hydrogen compartment, an oxygen compartment, and two electrodes. The cryogenic gas storage system, also located in the SM, supplies the hydrogen and oxygen used in the fuel cell power plants, as well as the oxygen used in the ECS.

Three silver-zinc oxide storage batteries supply power to the CM during entry and after landing, provide power for sequence controllers and supplement the fuel cells during periods of peak power demand. These batteries are located in the CM lower equipment bay. A battery charger is located in the same bay to assure a full charge prior to entry.

Two other silver-zinc oxide batteries, independent of and completely isolated from the rest of the dc power system, are used to supply power for explosive devices for CM/SM separation, parachute deployment and separation, third-stage separation, launch escape system tower separation, and other pyrotechnic uses.

Environmental Control System (ECS) — Controls spacecraft atmosphere, pressure, and temperature and manages water. In addition to regulating cabin and suit gas pressure, temperature and humidity, the system removes carbon dioxide, odors and particles, and ventilates the cabin after landing. It collects and stores fuel cell potable water for crew use, supplies water to the glycol evaporators for cooling, and dumps surplus water overboard through the urine dump valve. Proper operating temperature of electronics and electrical equipment is maintained by this system through the use of the cabin heat exchangers, the space radiators, and the glycol evaporators.

Recovery aids include the uprighting system, swimmer interphone connections, sea dye marker, flashing beacon, VHF recovery beacon, and VHF transceiver. The uprighting system consists of three compressor-inflated bags to upright the spacecraft if it should land in the water apex down (stable II position).

Caution and Warning System — Monitors spacecraft systems for out-of-tolerance conditions and alerts crew by visual and audible alarms so that crewmen may trouble-shoot the problem.

Controls and Displays — Provide readouts and control functions of all other spacecraft systems in the command and service modules. All controls are designed to be operated by crewmen in pressurized suits. Displays are grouped by system and located according to the frequency the crew refers to them.

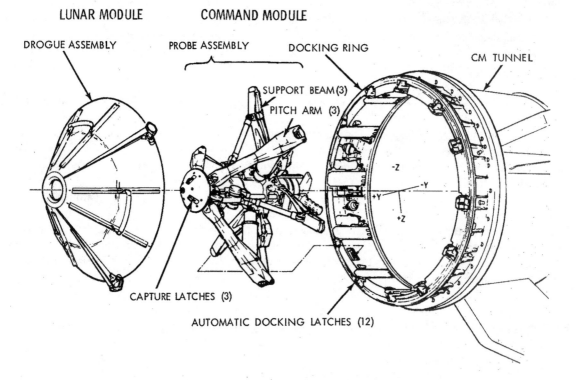

APOLLO DOCKING MECHANISMS

LUNAR MODULE STRUCTURES, WEIGHT

The lunar module is a two-stage vehicle designed for space operations near and on the Moon. The LM is incapable of reentering the atmosphere. The lunar module stands 22 feet 11 inches high and is 31 feet wide (diagonally across landing gear).

Joined by four explosive bolts and umbilicals, the ascent and descent stages of the LM operate as a unit until staging, when the ascent stage functions as a single spacecraft for rendezvous and docking with the CSM.

Ascent Stage

Three main sections make up the ascent stage — the crew compartment, midsection, and aft equipment bay. Only the crew compartment and midsection are pressurized (4.8 psig; 337.4 gm/sq cm) as part of the LM cabin; all other sections of the LM are unpressurized. The cabin volume is 235 cubic feet (6-7 cubic meters). The ascent stage measures 12 feet 4 inches high by 14 feet 1 inch in diameter.

Structurally, the ascent stage has six substructural areas: crew compartment, midsection, aft equipment bay, thrust chamber assembly cluster supports, antenna supports and thermal and micrometeoroid shield.

The cylindrical crew compartment is a semimonocoque structure of machined longerons and fusion-welded aluminum sheet and is 92 inches (2.35 m) in diameter and 42 inches (1.07 m) deep. Two flight stations are equipped with control and display panels, armrests, body restraints, landing aids, two front windows, an overhead docking window, and an alignment optical telescope in the center between the two flight stations. The habitable volume is 160 cubic feet.

Two triangular front windows and the 32-inch (0.81 m) square inward opening forward hatch are in the crew compartment front face.

External structural beams support the crew compartment and serve to support the lower interstage mounts at their lower ends. Ringstiffened semimonocoque construction is employed in the midsection, with chem-milled aluminum skin over fusion-welded longerons and stiffeners. Fore-and-aft beams across the top of the midsection join with those running across the top of the cabin to take all ascent stage stress loads and, in effect, isolate the cabin from stresses.

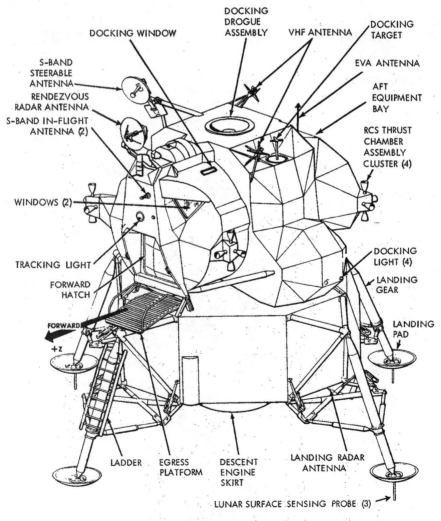

APOLLO LUNAR MODULE

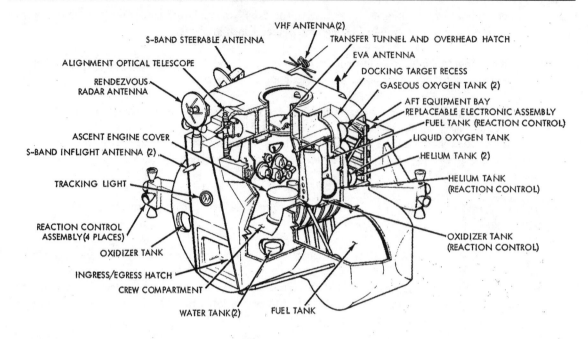

APOLLO LUNAR MODULE - ASCENT STAGE

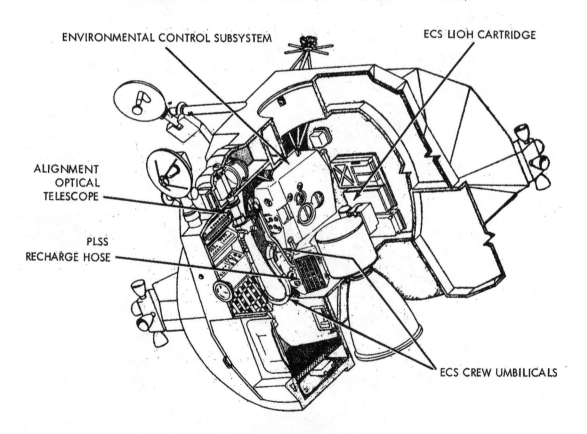

LM CABIN INTERIOR, LEFT HALF

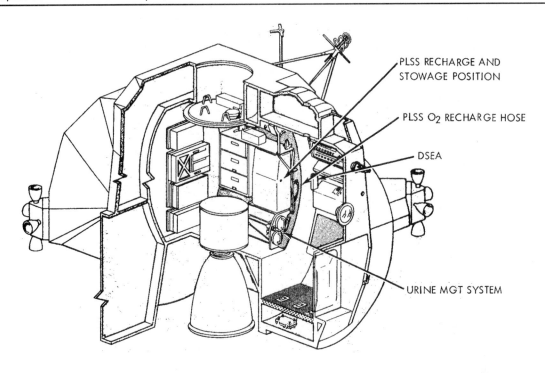

PLSS RECHARGE AND
STOWAGE POSITION

PLSS O$_2$ RECHARGE HOSE

DSEA

URINE MGT SYSTEM

LM CABIN INTERIOR, RIGHT HALF

The ascent stage engine compartment is formed by two beams running across the lower midsection deck and mated to the fore and aft bulkheads. Systems located in the midsection include the LM guidance computer, the power and servo assembly, ascent engine propellant tanks, RCS propellant tanks, the environmental control system, and the waste management section.

A tunnel ring atop the ascent stage meshes with the command module docking latch assemblies. During docking, the CM docking ring and latches are aligned by the LM drogue and the CSM probe.

The docking tunnel extends downward into the midsection 16 inches (40 cm). The tunnel is 32 inches (0.81 cm) in diameter and is used for crew transfer between the CSM and LM. The upper hatch on the inboard end of the docking tunnel hinges downward and cannot be opened with the LM pressurized and undocked.

A thermal and micrometeoroid shield of multiple layers of mylar and a single thickness of thin aluminum skin encases the entire ascent stage structure.

Descent Stage

The descent stage consists of a cruciform load-carrying structure of two pairs of parallel beams, upper and lower decks, and enclosure bulkheads — all of conventional skin-and-stringer aluminum alloy construction. The center compartment houses the descent engine, and descent propellant tanks are housed in the four square bays around the engine. The descent stage measures 10 feet 7 inches high by 14 feet 1 inch in diameter.

Four-legged truss outriggers mounted on the ends of each pair of beams serve as SLA attach points and as "knees" for the landing gear main struts.

Triangular bays between the main beams are enclosed into quadrants housing such components as the ECS water tank, helium tanks, descent engine control assembly of the guidance, navigation and control subsystem, ECS gaseous oxygen tank, and batteries for the electrical power system. Like the ascent stage, the descent stage is encased in the mylar and aluminum alloy thermal and micrometeoroid shield.

The LM external platform, or "porch", is mounted on the forward outrigger just below the forward hatch. A ladder extends down the forward landing gear strut from the porch for crew lunar surface operations.

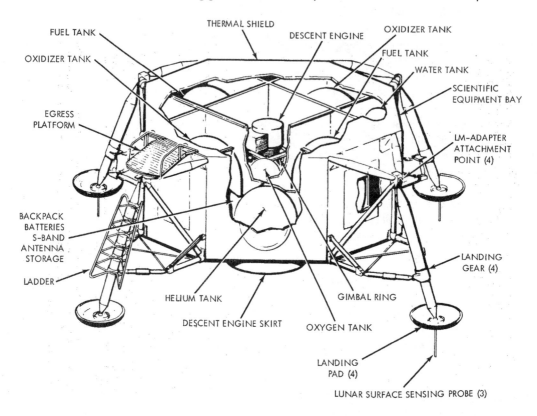

In a retracted position until after the crew mans the LM, the landing gear struts are explosively extended and provide lunar surface landing impact attenuation. The main struts are filled with crushable aluminum honeycomb for absorbing compression loads. Footpads 37 inches (0.95 m) in diameter at the end of each landing gear provide vehicle "floatation" on the lunar surface.

Each pad (except forward pad) is fitted with a lunar surface sensing probe which signals the crew to shut down the descent engine upon contact with the lunar surface.

LM-5 flown on the Apollo 11 mission will have a launch weight Of 33,205 pounds. The weight breakdown is as follows:

Ascent stage, dry	4,804 lbs.	Includes water and oxygen; no crew
Descent stage, dry	4,483 lbs.	
RCS propellants (loaded)	604 lbs.	
DPS propellants (loaded)	18,100 lbs.	
APS propellants (loaded)	5,214 lbs.	
Total	33,205 lbs.	

Lunar Module Systems

Electrical Power System — The LM DC electrical system consists of six silver zinc primary batteries — four in the descent stage and two in the ascent stage, each with its own electrical control assembly (ECA). Power feeders from all primary batteries pass through circuit breakers to energize the LM DC buses, from which 28-volt DC power is distributed through circuit breakers to all LM systems. AC power (117v 400Hz) is supplied by two inverters, either of which can supply spacecraft AC load needs to the AC buses.

Environmental Control System — Consists of the atmosphere revitalization section, oxygen supply and cabin pressure control section, water management, heat transport section, and outlets for oxygen and water servicing of the Portable Life Support System (PLSS).

Components of the atmosphere revitalization section are the suit circuit assembly which cools and ventilates the pressure garments, reduces carbon dioxide levels, removes odors, noxious gases and excessive moisture; the cabin recirculation assembly which ventilates and controls cabin atmosphere temperatures; and the steam flex duct which vents to space steam from the suit circuit water evaporator.

The oxygen supply and cabin pressure section supplies gaseous oxygen to the atmosphere revitalization section for maintaining suit and cabin pressure. The descent stage oxygen supply provides descent flight phase and lunar stay oxygen needs, and the ascent stage oxygen supply provides oxygen needs for the ascent and rendezvous flight phase.

Water for drinking, cooling, fire fighting, food preparation, and refilling the PLSS cooling water servicing tank is supplied by the water management section. The water is contained in three nitrogen pressurized bladder-type tanks, one of 367-pound capacity in the descent stage and two of 47.5-pound capacity in the ascent stage.

The heat transport section has primary and secondary water-glycol solution coolant loops. The primary coolant loop circulates waterglycol for temperature control of cabin and suit circuit oxygen and for thermal control of batteries and electronic components mounted on cold plates and rails. If the primary loop becomes inoperative, the secondary loop circulates coolant through the rails and cold plates only. Suit circuit cooling during secondary coolant loop operation is provided by the suit loop water boiler. Waste heat from both loops is vented overboard by water evaporation or sublimators.

Communication — Two S-band transmitter-receivers, two VHF transmitter-receivers, a signal processing assembly, and associated spacecraft antenna make up the LM communications system. The system transmits and receives voice, tracking and ranging data, and transmits telemetry data on about 270 measurements and TV signals to the ground. Voice communications between the LM and ground stations is by S-band, and between the LM and CSM voice is on VHF.

Although no real-time commands can be sent to LM-5 and subsequent spacecraft, the digital uplink is retained to process guidance officer commands transmitted from Mission Control Center to the LM guidance computer, such as state vector updates.

The data storage electronics assembly (DSEA) is a four channel voice recorder with timing signals with a 10-hour recording capacity which will be brought back into the CSM for return to Earth. DSEA recordings cannot be "dumped" to ground stations.

LM antennas are one 26-inch diameter parabolic S-band steerable antenna, two S-band inflight antennas, two VHF inflight antennas, and an erectable S-band antenna (optional) for lunar surface.

Guidance, Navigation and Control System — Comprised of six sections: primary guidance and navigation section (PGNS), abort guidance section (AGS), radar section, control electronics section (CES), and orbital rate drive electronics for Apollo and LM (ORDEAL).

* The PGNS is an aided inertial guidance system updated by the alignment optical telescope, an inertial measurement unit, and the rendezvous and landing radars. The system provides inertial reference data for computations, produces inertial alignment reference by feeding optical sighting data into the LM guidance computer, displays position and velocity data, computes LM-CSM rendezvous data from radar inputs, controls attitude and thrust to maintain desired LM trajectory, and controls descent engine throttling and gimbaling.

The LM-5 guidance computer has the Luminary IA software program for processing landing radar altitude and velocity information for lunar landing. LM-4, flown on Apollo 10, did not have the landing phase in its guidance computer Luminary I program.

* The AGS is an independent backup system for the PGNS, having its own inertial sensors and computer.

* The radar section is made up of the rendezvous radar which provides CSM range and range rate, and line-of-sight angles for maneuver computation to the LM guidance computer; the landing radar which provide altitude and velocity data to the LM guidance computer during lunar landing. The rendezvous radar has an operating range from 80 feet to 400 nautical miles. The range transfer tone assembly, utilizing VHF electronics, is a passive responder to the CSM VHF ranging device and is a backup to the rendezvous radar.

* The CES controls LM attitude and translation about all axes. It also controls by PGNS command the automatic operation of the ascent and descent engines, and the reaction control thrusters. Manual attitude controller and thrust-translation controller commands are also handled by the CES.

* ORDEAL, displays on the flight director attitude indicator, is the computed local vertical in the pitch axis during circular Earth or lunar orbits.

Reaction Control System — The LM has four RCS engine clusters of four 100 pound (45.4 kg) thrust engines each which use helium-pressurized hypergolic propellants. The oxidizer is nitrogen tetroxide, fuel is Aerozine 50 (50/50 blend of hydrazine and unsymmetrical dimethyl hydrazine). Propellant plumbing, valves and pressurizing components are in two parallel, independent systems, each feeding half the engines in each cluster. Either system is capable of maintaining attitude alone, but if one supply system fails, a propellant crossfeed allows one system to supply all 16 engines. Additionally, interconnect valves permit the RCS system to draw from ascent engine propellant tanks.

The engine clusters are mounted on outriggers 90 degrees apart on the ascent stage.

The RCS provides small stabilizing impulses during ascent and descent burns, controls LM attitude during maneuvers, and produces thrust for separation, and ascent/descent engine tank ullage. The system may be operated in either the pulse or steady-state modes.

Descent Propulsion System — Maximum rated thrust of the descent engine is 9,870 pounds (4,380.9 kg) and is throttleable between 1,050 pounds (476.7 kg) and 6,300 pounds (2,860.2 kg). The engine can be gimbaled six degrees in any direction in response to attitude commands and for offset center of gravity trimming. Propellants are helium-pressurized Aerozine 50 and nitrogen tetroxide.

Ascent Propulsion System — The 3,500-Pound (1,589 kg) thrust ascent engine is not gimbaled and performs at full thrust. The engine remains dormant until after the ascent stage separates from the descent stage. Propellants are the same as are burned by the RCS engines and the descent engine.

Caution and Warning controls and Displays — These two systems have the same function aboard the lunar module as they do aboard the command module. (See CSM systems section.)

Tracking and Docking Lights — A flashing tracking light (once per second, 20 millisecond duration) on the front face of the lunar module is an aid for contingency CSM-active rendezvous LM rescue. Visibility ranges from 400 nautical miles through the CSM sextant to 130 miles with the naked eye. Five docking lights analogous to aircraft running lights are mounted on the LM for CSM-active rendezvous: two forward yellow lights, aft white light, port red light and starboard green light. All docking lights have about a 1,000-foot visibility.

SATURN V LAUNCH VEHICLE DESCRIPTION AND OPERATION

The Apollo 11 spacecraft will be boosted into Earth orbit and then onto a lunar trajectory by the sixth Saturn V launch vehicle. The 281-foot high Saturn V generates enough thrust to place a 125-ton payload into a 105 nm Earth orbit or boost about 50 tons to lunar orbit.

The Saturn V, developed by the NASA-Marshall Space Flight Center, underwent research and development testing in the "all-up" mode. From the first launch all stages have been live. This has resulted in "man rating" of the Saturn V in two launches. The third Saturn V (AS-503) carried Apollo 8 and its crew on a lunar orbit mission.

Saturn V rockets were launched November 9, 1967, April 4, 1968, December 21, 1968, March 3, 1969, and May 18, 1969. The first two space vehicle were unmanned; the last three carried the Apollo 8, 9 and 10 crews, respectively.

Launch Vehicle Range Safety Provisions

In the event of an imminent emergency during the launch vehicle powered flight phase it could become necessary to abort the mission and remove the command module and crew from immediate danger. After providing for crew safety, the Range Safety Officer may take further action if the remaining intact vehicle constitutes a hazard to overflown geographic areas. Each launch vehicle propulsive stage is equipped with a propellant dispersion system to terminate the vehicle flight in a safe location and disperse propellants with a minimized ignition probability. A transmitted ground command shuts down all engines and a second command detonates explosives which open the fuel and oxidizer tanks enabling the propellants to disperse. On each stage the tank cuts are made in non-adjacent areas to minimize propellant mixing. The stage propellant dispersion systems are safed by ground command.

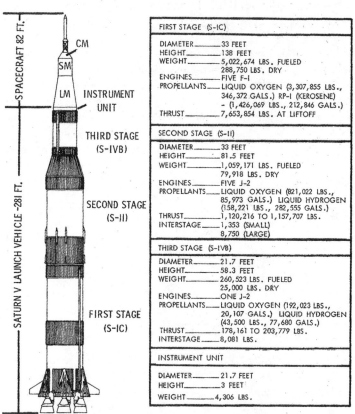

SATURN V LAUNCH VEHICLE

FIRST STAGE (S-IC)	
DIAMETER	33 FEET
HEIGHT	138 FEET
WEIGHT	5,022,674 LBS. FUELED
	288,750 LBS. DRY
ENGINES	FIVE F-1
PROPELLANTS	LIQUID OXYGEN (3,307,855 LBS.,
	346,372 GALS.) RP-1 (KEROSENE)
	- (1,426,069 LBS., 212,846 GALS.)
THRUST	7,653,854 LBS. AT LIFTOFF

SECOND STAGE (S-II)	
DIAMETER	33 FEET
HEIGHT	81.5 FEET
WEIGHT	1,059,171 LBS. FUELED
	79,918 LBS. DRY
ENGINES	FIVE J-2
PROPELLANTS	LIQUID OXYGEN (821,022 LBS.,
	85,973 GALS.) LIQUID HYDROGEN
	(158,221 LBS., 282,555 GALS.)
THRUST	1,120,216 TO 1,157,707 LBS.
INTERSTAGE	1,353 (SMALL)
	8,750 (LARGE)

THIRD STAGE (S-IVB)	
DIAMETER	21.7 FEET
HEIGHT	58.3 FEET
WEIGHT	260,523 LBS. FUELED
	25,000 LBS. DRY
ENGINES	ONE J-2
PROPELLANTS	LIQUID OXYGEN (192,023 LBS.,
	20,107 GALS.) LIQUID HYDROGEN
	(43,500 LBS., 77,680 GALS.)
THRUST	178,161 TO 203,779 LBS.
INTERSTAGE	8,081 LBS.

INSTRUMENT UNIT	
DIAMETER	21.7 FEET
HEIGHT	3 FEET
WEIGHT	4,306 LBS.

NOTE: WEIGHTS AND MEASURES GIVEN ABOVE ARE FOR THE NOMINAL VEHICLE CONFIGURATION FOR APOLLO 11. THE FIGURES MAY VARY SLIGHTLY DUE TO CHANGES BEFORE LAUNCH TO MEET CHANGING CONDITIONS. WEIGHTS NOT INCLUDED IN ABOVE ARE FROST AND MISCELLANEOUS SMALLER ITEMS.

SPACE VEHICLE WEIGHT SUMMARY (pounds)

Event	Wt. Chg.	Veh. Wt.
At ignition		6,484,280
Thrust buildup propellant used	85,745	
At first motion		6,398,535
S-IC frost	650	
S-IC nitrogen purge	37	
S-II frost	450	
S-II insulation purge gas	120	
S-IVB frost	200	
Center engine decay propellant used	2,029	
Center engine expended propellant	406	
S-IC mainstage propellant used	4,567,690	
Outboard engine decay propellant used	8,084	
S-IC stage drop weight	363,425	
S-IC/S-II small interstage	1,353	
S-II ullage propellant used	73	
At S-IC separation		1,454,014
S-II thrust buildup propellant used	1,303	
S-II start tank	25	
S-II ullage propellant used	1,288	
S-II mainstage propellant and venting	963,913	
Launch escape tower	8,930	
S-II aft interstage	8,750	
S-II thrust decay propellant used	480	
S-II stage drop weight	94,140	
S-II/S-IVB interstage	8,081	
S-IVB aft frame dropped	48	
S-IVB detonator package	3	
At S-II/S-IVB separation		367,053
S-IVB ullage rocket propellant	96	
At S-IVB ignition		366,957
S-IVB ullage propellant	22	
S-IVB hydrogen in start tank	4	
Thrust buildup propellant	436	
S-IVB mainstage propellant used	66,796	
S-IVB ullage rocket cases	135	
S-IVB APS propellant	2	
At first S-IVB cutoff signal		299,586
Thrust decay propellant used	89	
APS propellant (ullage)	5	
Engine propellant lost	30	
At parking orbit insertion		299,562
Fuel tank vent	2,879	
APS propellant	235	
Hydrogen in start tank	2	
O2/H2 burner	16	
LOX tank vent	46	
S-IVB fuel lead loss	5	
At second S-IVB ignition		296,379
S-IVB hydrogen in start tank	4	
Thrust buildup propellant	569	
S-IVB mainstage propellant used	164,431	
APS propellant used	8	
At second S-IVB cutoff signal		139,533
Thrust decay propellant used	124	
Engine propellant lost	40	
At translunar injection		139,369

First Stage

The 7.6 million pound thrust first stage (S-IC) was developed jointly by the National Aeronautics and Space Administration's Marshall Space Flight Center and the Boeing Co.

The Marshall Center assembled four S-IC stages: a structural test model, a static test version, and the first two flight stages. Subsequent flight stages are assembled by Boeing at the Michoud Assembly Facility, New Orleans.

The S-IC for the Apollo 11 mission was the third flight booster tested at the NASA-Mississippi Test Facility. the first S-IC test at MTF was on May 11, 1967, the second on August 9, 1967, and the third— the booster for Apollo 11— was on August 6, 1968. Earlier flight stages were static fired at the Marshall Center.

The booster stage stands 138 feet high and is 33 feet in diameter. Major structural components include thrust structure, fuel tank, intertank structure, oxidizer tank, and forward skirt. Its five engines burn kerosene (RP-1) fuel and liquid oxygen. The stage weighs 288,750 empty and 5,022,674 pounds fueled.

Normal propellant flow rate to the five F-1 engines is 29,364.5 pounds (2,230 gallons) per second. Four of the engines are mounted on a ring, at 90 degree intervals. These four are gimballed to control the rocket's direction of flight. The fifth engine is mounted rigidly in the center.

Second Stage

The Space Division of North American Rockwell Corp. builds the 1 million pound thrust S-II stage at Seal Beach, California. The 81 foot 7 inch long, 33 foot diameter stage is made up of the forward skirt to which the third stage attaches, the liquid hydrogen tank, liquid oxygen tank (separated from the hydrogen tank by an insulated common bulkhead), the thrust structure on which the engines are mounted, and an interstage section to which the first stage attaches.

Five J-2 engines power the S-II. The outer four engines are equally spaced on a 17.5 foot diameter circle. These four engines may be gimballed through a plus or minus seven-degree square pattern for thrust vector control. As on the first stage, the center engine (number 5) is mounted on the stage centerline and is fixed in position.

The second stage (S-II), like the third stage, uses high performance J-2 engines that burn liquid oxygen and liquid hydrogen. The stage's purpose is to provide stage boost almost to Earth orbit.

The S-II for Apollo 11 was static tested by North American Rockwell at the NASA-Mississippi Test Facility on September 3, 1968. This stage was shipped to test site via the Panama Canal for the test firing.

Third Stage

The third stage (S-IVB) was developed by the McDonnell Douglas Astronautics Co. at Huntington Beach, Calif. At Sacramento, Calif., the stage passed a static firing test on July 17, 1968, as part of Apollo 11 mission preparation. The stage was flown directly to the NASA-Kennedy Space Center by the special aircraft, Super Guppy.

Measuring 58 feet 4 inches long and 21 feet 8 inches in diameter, the S-IVB weighs 25,000 pounds dry. At first ignition, it weighs 262,000 pounds. The interstage section weighs an additional 8,081 pounds.

The fuel tanks contain 43,500 pounds of liquid hydrogen and 192,023 pounds of liquid oxygen at first ignition, totalling 235,523 pounds of propellants. Insulation between the two tanks is necessary because the liquid oxygen, at about 293 degrees below zero Fahrenheit, is warm enough, relatively, to rapidly heat the liquid hydrogen, at 423 degrees below zero, and cause it to turn to gas. The single J-2 engine produces a maximum 230,000 pounds of thrust. The stage provides propulsion twice during the Apollo 11 mission.

Instrument Unit

The instrument unit (IU) is a cylinder three feet high and 21 feet 8 inches in diameter. It weighs 4,306 pounds and contains the guidance, navigation and control equipment to steer the vehicle through its Earth orbits and into the final translunar injection maneuver.

The IU also contains telemetry, communications, tracking, and crew safety systems, along with its own supporting electrical power and environmental control systems.

Components making up the "brain" of the Saturn V are mounted on cooling panels fastened to the inside surface of the instrument unit skin. The "cold plates" are part of a system that removes heat by circulating cooled fluid through a heat exchanger that evaporates water from a separate supply into the vacuum of space.

The six major systems of the instrument unit are structural, thermal control, guidance and control, measuring and telemetry, radio frequency, and electrical.

The instrument unit provides navigation, guidance, and control of the vehicle measurement of the vehicle performance and environment; data transmission with ground stations; radio tracking of the vehicle: checkout and monitoring of vehicle functions; initiation of stage functional sequencing; detection of emergency situations; generation and network distribution of electric power system operation; and preflight checkout and launch and flight operations.

A path-adaptive guidance scheme is used in the Saturn V instrument unit. A programmed trajectory is used during first stage boost with guidance beginning only after the vehicle has left the atmosphere. This is to prevent movements that might cause the vehicle to break apart while attempting to compensate for winds, jet streams, and gusts encountered in the atmosphere.

If after second stage ignition the vehicle deviates from the optimum trajectory in climb, the vehicle derives and corrects to a new trajectory. Calculations are made about once each second throughout the flight. The launch vehicle digital computer and data adapter perform the navigation and guidance computations and the flight control computer converts generated attitude errors into control commands.

The ST-124M inertial platform— the heart of the navigation, guidance and control system— provides space-fixed reference coordinates and measures acceleration along the three mutually perpendicular axes of the coordinate system. If the inertial platform fails during boost, spacecraft systems continue guidance and control functions for the rocket. After second stage ignition the crew can manually steer the space vehicle.

International Business Machines Corp., is prime contractor for the instrument unit and is the supplier of the guidance signal processor and guidance computer. Major suppliers of instrument unit components are: Electronic Communications, Inc., control computer; Bendix Corp., ST-124M inertial platform; and IBM Federal Systems Division, launch vehicle digital computer and launch vehicle data adapter.

Propulsion

The 41 rocket engines of the Saturn V have thrust ratings ranging from 72 pounds to more than 1.5 million pounds. Some engines burn liquid propellants, others use solids.

The five F-1 engines in the first stage burn RP-1 (kerosene) and liquid oxygen. Engines in the first stage develop approximately 1,530,771 pounds of thrust each at liftoff, building up to about 1,817,684 pounds before cutoff. The cluster of five engines gives the first stage a thrust range of from 7,653,854 pounds at liftoff to 9,088,419 pounds just before center engine cutoff.

The F-1 engine weighs almost 10 tons, is more than 18 feet high and has a nozzle-exit diameter of nearly 14 feet. The F-1 undergoes static testing for an average 650 seconds in qualifying for the 160 second run during the Saturn V first stage booster phase. The engine consumes almost three tons of propellants per second.

The first stage of the Saturn V for this mission has eight other rocket motors. These are the solid-fuel retrorockets which will slow and separate the stage from the second stage. Each rocket produces a thrust of 87,900 pounds for 0.6 second.

The main propulsion for the second stage is a cluster of five J-2 engines burning liquid hydrogen and liquid oxygen. Each engine develops a mean thrust of more than 227,000 pounds at 5.1 mixture ratio (variable from 224,000 to 231,000 in phases of this flight), giving the stage a total mean thrust of more than 1.135 million pounds.

Designed to operate in the hard vacuum of space, the 3,500 pound J-2 is more efficient than the F-1 because it burns the high-energy fuel hydrogen. F-1 and J-2 engines are produced by the Rocketdyne Division of North American Rockwell Corp.

The second stage has four 21,000-pound-thrust solid fuel rocket engines. These are the ullage rockets mounted on the S-IC/S-II interstage section. These rockets fire to settle liquid propellant in the bottom of the main tanks and help attain a "clean" separation from the first stage; they remain with the interstage when it drops away at second plane separation. Four retrorockets are located in the S-IVB aft interstage (which never separates from the S-II) to separate the S-II from the S-IVB prior to S-IVB ignition.

Eleven rocket engines perform various functions on the third stage. A single J-2 provides the main propulsive force; there are two jettisonable main ullage rockets and eight smaller engines in the two auxiliary propulsion system modules.

Launch Vehicle Instrumentation and Communication

A total of 1,348 measurements will be taken in flight on the Saturn V launch vehicle: 330 on the first stage, 514 on the second stage, 283 on the third stage, and 221 on the instrument unit.

Telemetry on the Saturn V includes FM and PCM systems on the S-IC, two FM and a PCM on the S-II, a PCM on the S-IVB, and an FM, a PCM and a CCS on the IU. Each propulsive stage has a range safety system, and the IU has C-Band and command systems.

Note: FM (Frequency Modulated) PCM (Pulse Code Modulated) CCS (Command Communications System)

S-IVB Restart

The third stage of the Saturn V rocket for the Apollo mission will burn twice in space. The second burn places the spacecraft on the translunar trajectory. The first opportunity for this burn is at 2 hours 44 minutes and 15 seconds after launch.

The primary pressurization system of the propellant tanks for the S-IVB restart uses a helium heater. In this system, nine helium storage spheres in the liquid hydrogen tank contain gaseous helium charged to about 3,000 psi. This helium is passed through the heater which heats and expands the gas before it enters the propellant tanks. The heater operates on hydrogen and oxygen gas from the main propellant tanks.

The backup system consists of five ambient helium spheres mounted on the stage thrust structure. This system, controlled by the fuel repressurization control module, can repressurize the tanks in case the primary system fails. The restart will use the primary system. If that system fails, the backup system will be used.

Differences in Launch Vehicles for Apollo 10 and Apollo 11

The greatest difference between the Saturn V launch vehicle for Apollo 10 and the one for Apollo 11 is in the number of instrumentation measurements planned for the flight. Apollo 11 will be flying the operational configuration of instrumentation. Most research and development instrumentation has been removed, reducing the total number of measurements from 2,342 on Apollo 10 to 1,348 on Apollo 11. Measurements on Apollo 10, with Apollo 11 measurements in parentheses, were: S-IC 672 (330); S-II, 980 (514); S-IVB, 386 (283); and IU, 298 (221).

The center engine of the S-II will be cut off early, as was done during the Apollo 10 flight, to eliminate the longitudinal oscillations reported by astronauts on the Apollo 9 mission. Cutting off the engine early on Apollo 10 was the simplest and quickest method of solving the problem.

APOLLO 11 CREW

Life Support Equipment - Space Suits

Apollo 11 crewmen will wear two versions of the Apollo space suit: an intravehicular pressure garment assembly worn by the command module pilot and the extravehicular pressure garment assembly worn by the commander and the lunar module pilot. Both versions are basically identical except that the extravehicular version has an integral thermal/ meteoroid garment over the basic suit.

From the skin out, the basic pressure garment consists of a nomex comfort layer, a neoprene-coated nylon pressure bladder and a nylon restraint layer. The outer layers of the intravehicular suit are, from the inside out, nomex and two layers of Teflon-coated Beta cloth. The extravehicular integral thermal/meteoroid cover consists of a liner of two layers of neoprene-coated nylon, seven layers of Beta/kapton spacer laminate, and an outer layer of Teflon-coated Beta fabric.

The extravehicular suit together with a liquid cooling garment, portable life support system (PLSS), oxygen purge system, lunar extravehicular visor assembly and other components make up the extravehicular mobility unit (EMU). The EMU provides an extravehicular crewman with life support for a four hour mission outside the lunar module without replenishing expendables. EMU total weight is 183 pounds. The intravehicular suit weighs 35.6 pounds.

Liquid cooling garment— A knitted nylon-spandex garment with a network of plastic tubing through which cooling water from the PLSS is circulated. It is worn next to the skin and replaces the constant wear-garment during EVA only.

Portable Life Support System — A backpack supplying oxygen at 3.9 psi and cooling water to the liquid cooling garment. Return oxygen is cleansed of solid and gas contaminants by a lithium hydroxide canister. The PLSS includes communications and telemetry equipment, displays and controls, and a main power supply. The PLSS is covered by a thermal insulation jacket. (Two stowed in LM).

Oxygen Purge System — Mounted atop the PLSS, the oxygen purge system provides a contingency 30-minute supply of gaseous oxygen in two two-pound bottles pressurized to 5,880 psia. The system may also be worn separately on the front of the pressure garment assembly torso. It serves as a mount for the VHF antenna for the PLSS. (Two stowed in LM).

Lunar extravehicular visor assembly— A polycarbonate shell and two visors with thermal control and optical coatings on them. The EVA visor is attached over the pressure helmet to provide impact, micrometeoroid, thermal and ultraviolet infrared light protection to the EVA crewman.

Extravehicular gloves— Built of an outer shell of Chromel-R fabric and thermal insulation to provide protection when handling extremely hot and cold objects. The finger tips are made of silicone rubber to provide the crewman more sensitivity.

A one-piece constant-wear garment, similar to "long Johns", is worn as an undergarment for the spacesuit in intravehicular operations and for the inflight coveralls. The garment is porous-knit cotton with a waist--to-neck zipper for donning. Biomedical harness attach, points are provided.

During periods out of the space suits, crewmen will wear two-piece Teflon fabric inflight coveralls for warmth and for pocket stowage of personal Items.

Communications carriers ("Snoopy hats") with redundant microphones and earphones are worn with the pressure helmet; a lightweight headset is worn with the inflight coveralls.

CONNECTOR

MANIFOLD

ZIPPER

TYGON TUBING

DOSIMETER

LIQUID COOLING GARMENT

EXTRAVEHICULAR MOBILITY UNIT

BACKPACK SUPPORT STRAPS

OXYGEN PURGE SYSTEM

LUNAR EXTRAVEHICULAR VISOR

BACKPACK CONTROL BOX

SUNGLASSES POCKET

OXYGEN PURGE SYSTEM ACTUATOR

BACKPACK

PENLIGHT POCKET

CONNECTOR COVER

COMMUNICATION, VENTILATION, AND LIQUID COOLING UMBILICALS

OXYGEN PURGE SYSTEM UMBILICAL

LM RESTRAINT RING

INTEGRATED THERMAL METEOROID GARMENT

URINE TRANSFER CONNECTOR, BIOMEDICAL INJECTION, DOSIMETER ACCESS FLAP AND DONNING LANYARD POCKET

EXTRAVEHICULAR GLOVE

UTILITY POCKET

LUNAR OVERSHOE

72" MAXIMUM REACH HEIGHT

66" MAXIMUM WORKING HEIGHT

48"

OPTIMUM WORKING HEIGHT

30"

28" MINIMUM WORKING HEIGHT

22" MINIMUM REACH HEIGHT

ASTRONAUT REACH CONSTRAINTS

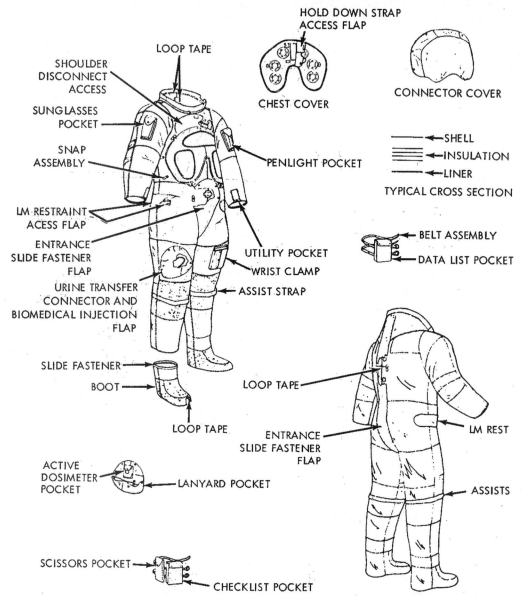

INTEGRATED THERMAL MICROMETEROID GARMENT

APOLLO 11 CREW MENU

The Apollo 11 crew had a wide range of food items from which to select their daily mission space menu. More than 70 items comprise the food selection list of freeze-dried rehydratable, wetpack and spoon-bowl foods.

Balanced meals for five days have been packed in man/day overwraps, and items similar to those in the daily menus have been packed in a sort of snack pantry. The snack pantry permits the crew to locate easily a food item in a smorgasbord mode without having to "rob" a regular meal somewhere down deep in a storage box.

Water for drinking and rehydrating food is obtained from three sources in the command module— a dispenser for drinking water and two water spigots at the food preparation station, one supplying water at about 155 degrees F, the other at about 55 degrees F. The potable water dispenser squirts water continuously as long as the trigger is held down, and the food preparation spigots dispense water in one-ounce increments. Command module potable water is supplied from service module fuel cell by-product water.

A continuous-feed hand water dispenser similar to the one in the command module is used aboard the lunar module for cold-water rehydration of food packets stowed aboard the LM.

After water has been injected into a food bag, it is kneaded for about three minutes. The bag neck is then cut off and the food squeezed into the crewman's mouth. After a meal, germicide pills attached to the outside of the food bags are placed in the bags to prevent fermentation and gas formation. The bags are then rolled and stowed in waste disposal compartments.

The day-by-day, meal-by-meal Apollo 11 menu for each crewman as well as contents of the snack pantry are listed on the following pages:

APOLLO XI (ARMSTRONG)

MEAL DAY 1*, 5	DAY 2	DAY 3	DAY 4
A	Peaches	Fruit Cocktail	Peaches	Canadian Bacon and Applesauce
	Bacon Squares (8)	Sausage Patties**	Bacon Squares (8)	Sugar Coated Corn Flakes
	Strawberry Cubes (4)		Cinn. Tstd. Bread Cubes (4)	Apricot Cereal Cubes (4)
			Peanut Cubes (4)	
	Grape Drink	Cocoa	Grape Drink	Cocoa
	Orange Drink	Grapefruit Drink	Orange Drink	Orange-Grapefruit Drink
B	Beef and Potatoes***	Frankfurters***	Cream of Chicken Soup	Shrimp Cocktail
	Butterscotch Pudding	Applesauce	Turkey and Gravy***	Ham and Potatoes***
	Brownies (4)	Chocolate Pudding	Cheese Cracker Cubes (6)	Fruit Cocktail
	Grape Punch	Orange-Grapefruit Drink	Chocolate Cubes (6)	Date Fruitcake (4)
		Pineapple-Grapefruit Drink	Grapefruit Drink	
C	Salmon Salad	Spaghetti with Meat Sauce**	Tuna Salad	Beef Stew**
	Chicken and Rice**	Pork and Scalloped Potatoes**	Chicken Stew**	Coconut Cubes (4)
	Sugar Cookie Cubes (6)	Pineapple Fruitcake (4)	Butterscotch Pudding	Banana Pudding
	Cocoa	Grape Punch	Cocoa	Grape Punch
	Pineapple-Grapefruit Drink		Grapefruit Drink	

APOLLO XI (COLLINS)

MEAL	DAY 1*, 5	DAY 2	DAY 3	DAY 4
A	Peaches	Fruit Cocktail	Peaches	Canadian Bacon and Applesauce
	Bacon Squares (8)	Sausage Patties**	Bacon Squares (8)	Sugar Coated Corn Flakes
	Strawberry Cubes (4)	Cinn. Tstd. Bread Cubes (4)	Apricot Cereal Cubes (4)	Peanut Cubes (4)
	Grape Drink	Cocoa	Grape Drink	Cocoa
	Orange Drink	Grapefruit Drink	Orange Drink	Orange-Grapefruit Drink
B	Beef and Potatoes***	Frankfurters***	Cream of Chicken Soup	Shrimp Cocktail
	Butterscotch Pudding	Applesauce	Turkey and Gravy***	Ham and Potatoes***
	Brownies (4)	Chocolate Pudding	Cheese Cracker Cubes (6)	Fruit Cocktail
	Grape Punch	Orange-Grapefruit Drink	Chocolate Cubes (4)	Date Fruitcake (4)
		Pineapple-Grapefruit Drink	Grapefruit Drink	
C	Salmon Salad	Potato Soup	Tuna Salad	Beef Stew**
	Chicken and Rice**	Pork and Scalloped Potatoes**	Chicken Stew**	Coconut Cubes (4)
	Sugar Cookie Cubes (6)	Pineapple Fruitcake (4)	Butterscotch Pudding	Banana Pudding
	Cocoa	Grape Punch	Cocoa	Grape Punch
	Pineapple-Grapefruit Drink		Grapefruit Drink	

APOLLO XI (ALDRIN)

MEAL DAY 1*, 5	DAY 2	DAY 3	DAY 4
A Peaches Bacon Squares (8) Strawberry Cubes (4) Grape Drink Orange Drink	Fruit Cocktail Sausage Patties** Cinn.Tstd. Bread Cubes (4) Cocoa Grapefruit Drink	Peaches Bacon Squares (8) Apricot Cereal Cubes (4) Grape Drink Orange Drink	Canadian Bacon and Applesauce Sugar Coated Corn Flakes Peanut Cubes (4) Cocoa Orange-Grapefruit Drink
B Beef and Potatoes*** Butterscotch Pudding Brownies (4) Grape Punch	Frankfurters*** Applesauce Chocolate Pudding Orange-Grapefruit Drink Pineapple-Grapefruit Drink	Cream of Chicken Soup Turkey and Gravy*** Cheese Cracker Cubes (6) Chocolate Cubes (6) Grapefruit Drink	Shrimp Cocktail Ham and Potatoes*** Fruit Cocktail Date Fruitcake (4)
C Salmon Salad Chicken and Rice** Sugar Cookie Cubes (4) Cocoa Pineapple-Grapefruit Drink	Chicken Salad Chicken and Gravy Beef Sandwiches (6) Pineapple Fruitcake (4) Grape Punch	Tuna Salad Chicken Stew** Butterscotch Pudding Cocoa Grapefruit Drink	Pork and Scalloped Potatoes** Coconut Cubes (4) Banana Pudding Grape Punch

*Day 1 consists of Meal B and C only **Spoon-Bowl Package ***Wet-Pack Food

ACCESSORIES	Unit
Chewing gum	15
Wet skin cleaning towels	30
Oral Hygiene Kit	1
3 toothbrushes	
1 edible toothpaste	
1 dental floss	
Contingency Feeding System	1
3 food restrainer pouches	
3 beverage packages	
1 valve adapter (pontube)	
Spoons	3

Snack Pantry

Breakfast	Units
Peaches	6
Fruit Cocktail	6
Canadian Bacon and Applesauce	3
Bacon Squares (8)	12
Sausage Patties*	3
Sugar Coated Corn Flakes	6
Strawberry Cubes (4)	3
Cinn.Tstd. Bread Cubes (4)	6
Apricot Cereal Cubes (4)	3
Peanut Cubes (4)	3
Total	51

Rehydratable Desserts	Units
Banana Pudding	6
Butterscotch Pudding	6
Applesauce	6
Chocolate Pudding	6
Total	24

Dried Fruits	Units	Stow
Apricots	6	1
Peaches	6	1
Pears	6	1

Salads/Meats	
Salmon Salad	3
Tuna Salad	3
Cream of Chicken Soup	6
Shrimp Cocktail	6
Spaghetti and Meat Sauce*	6
Beef Pot Roast	3
Beef and Vegetables	3
Chicken and Rice*	6

	Units	Stow
Chicken Stew*	3	
Beef Stew*	3	
Pork and Scalloped Potatoes*	6	
Ham and Potatoes (Wet)	3	
Turkey and Gravy (Wet)	6	
Total	57	

Beverages		
Orange Drink	6	
Orange-Grapefruit Drink	3	
Pineapple-Grapefruit Drink	3	
Grapefruit Drink	3	
Grape Drink	6	
Grape Punch	3	
Cocoa	6	
Coffee (B)	15	
Coffee (S)	15	
Coffee (C and S)	15	
Total	75	

Sandwich Spread	Units	Stow
Ham Salad (5 oz.)	1	1
Tuna Salad (5 oz.)	1	1
Chicken Salad (5 oz.)	1	1
Cheddar Cheese (2 oz.)	3	1

Bread		
Rye	6	6
White	6	6

Bites		
Cheese Cracker Cubes (6)	6	
BBQ Beef Bits (4)	6	
Chocolate Cubes (4)	6	
Brownies (4)	6	
Date Fruitcake (4)	6	
Pineapple Fruitcake (4)	6	
Jellied Fruit Candy (4)	6	
Caramel Candy (4)	6	

*Spoon-Bowl Package

LM-5 Food

Meal A.	Bacon Squares(8)
	Peaches
	Sugar Cookie Cubes (6)
	Coffee
	Pineapple-Grapefruit drink

Meal B.	Beef stew
	Cream of Chicken Soup
	Date Fruit Cake (4)
	Grape Punch
	Orange Drink

		Units
Extra Beverage		8
Dried Fruit		4
Candy Bar		4
Bread		2
Ham Salad Spread (tube food)	I	
Turkey and Gravy		2
Spoons		2

Personal Hygiene

Crew personal hygiene equipment aboard Apollo 11 includes body cleanliness items, the waste management system and one medical kit.

Packaged with the food are a toothbrush and a two-ounce tube of toothpaste for each crewman. Each man-meal package contains a 3.5-by-4 inch wet-wipe cleansing towel. Additionally, three packages of 12-by-12-inch dry towels are stowed beneath the command module pilot's couch. Each package contains seven towels. Also stowed under the command module pilot's couch are seven tissue dispensers containing 53 three-ply tissues each.

Solid body wastes are collected in Gemini-type plastic defecation bags which contain a germicide to prevent bacteria and gas formation. The bags are sealed after use and stowed in empty food containers for post-flight analysis.

Urine collection devices are provided for use while wearing either the pressure suit or the inflight coveralls. The urine is dumped overboard through the spacecraft urine dump valve in the CM and stored in the LM.

Medical Kit

The 5x5x8-inch medical accessory kit is stowed in a compartment on the spacecraft right side wall beside the lunar module pilot couch. The medical kit contains three motion sickness injectors, three pain suppression injectors, one two-ounce bottle first aid ointment, two one-ounce bottle eye drops, three nasal sprays, two compress bandages, 12 adhesive bandages, one oral thermometer and four spare crew biomedical harnesses. Pills in the medical kit are 60 antibiotic, 12 nausea, 12 stimulant, 18 pain killer, 60 decongestant, 24 diarrhea, 72 aspirin and 21 sleeping. Additionally, a small medical kit containing four stimulant, eight diarrhea, two sleeping and four pain killer pills, 12 aspirin, one bottle eye drops and two compress bandages is stowed in the lunar module flight data file compartment.

Survival Gear

The survival kit is stowed in two rucksacks in the right-hand forward equipment bay above the lunar module pilot.

Contents of rucksack No. 1 are: two combination survival lights, one desalter kit, three pair sunglasses, one radio beacon, one spare radio beacon battery and spacecraft connector cable, one knife in sheath, three water containers and two containers of Sun lotion.

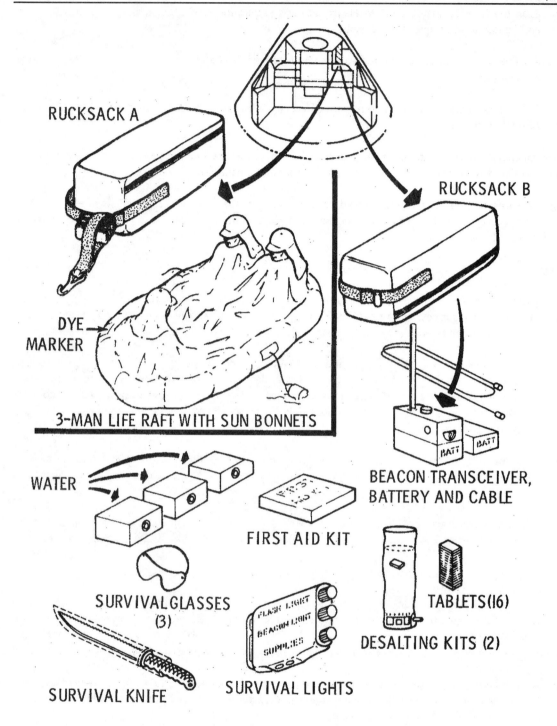

RUCKSACK A

RUCKSACK B

DYE MARKER

3-MAN LIFE RAFT WITH SUN BONNETS

WATER

FIRST AID KIT

BEACON TRANSCEIVER, BATTERY AND CABLE

SURVIVAL GLASSES (3)

SURVIVAL KNIFE

SURVIVAL LIGHTS

TABLETS (16)

DESALTING KITS (2)

Rucksack No. 2: one three-man life raft with CO2 inflater, one sea anchor, two sea dye markers, three sunbonnets, one mooring lanyard, three manlines, and two attach brackets.

The survival kit is designed to provide a 48-hour postlanding (water or land) survival capability for three crewmen between 40 degrees North and South latitudes.

Biomedical Inflight Monitoring

The Apollo 11 crew biomedical telemetry data received by the Manned Space Flight Network will be relayed for instantaneous display at Mission Control Center where heart rate and breathing rate data will be displayed on the flight surgeon's console. Heart rate and respiration rate average, range and deviation are computed and displayed on digital TV screens.

In addition, the instantaneous heart rate, real-time and delayed EKG and respiration are recorded on strip charts for each man.

Biomedical telemetry will be simultaneous from all crewmen while in the CSM, but selectable by a manual onboard switch in the LM.

Biomedical data observed by the flight surgeon and his team in the Life Support Systems Staff Support Room will be correlated with spacecraft and space suit environmental data displays.

Blood pressures are no longer telemetered as they were in the Mercury and Gemini programs. Oral temperature, however, can be measured onboard for diagnostic purposes and voiced down by the crew in case of inflight illness.

Training

The crewmen of Apollo 11 have spent more than five hours of formal crew training for each hour of the lunar-orbit mission's eight-day duration. More than 1,000 hours of training were in the Apollo 11 crew training syllabus over and above the normal preparations for the mission-technical briefings and reviews, pilot meetings and study.

The Apollo 11 crewmen also took part in spacecraft manufacturing checkouts at the North American Rockwell plant in Downey, Calif., at Grumman Aircraft Engineering Corp., Bethpage, N.Y., and in prelaunch testing at NASA Kennedy Space Center. Taking part in factory and launch area testing has provided the crew with thorough operational knowledge of the complex vehicle.

Highlights of specialized Apollo 11 crew training topics are:

Detailed series of briefings on spacecraft systems, operation and modifications.

* Saturn launch vehicle briefings on countdown, range safety, flight dynamics, failure modes and abort conditions. The launch vehicle briefings were updated periodically.

* Apollo Guidance and Navigation system briefings at the Massachusetts Institute of Technology Instrumentation Laboratory.

* Briefings and continuous training on mission photographic objectives and use of camera equipment.

Extensive pilot participation in reviews of all flight procedures for normal as well as emergency situations.

* Stowage reviews and practice in training sessions in the spacecraft, mockups and command module simulators allowed the crewmen to evaluate spacecraft stowage of crew-associated equipment.

* More than 400 hours of training per man in command module and lunar module simulators at MSC and KSC, including closed-loop simulations with flight controllers in the Mission Control Center. Other Apollo simulators at various locations were used extensively for specialized crew training.

* Entry corridor deceleration profiles at lunar-return conditions in the MSC Flight Acceleration Facility manned centrifuge.

* Lunar surface briefings and 1-g walk-throughs of lunar surface EVA operations covering lunar geology and microbiology and deployment of experiments in the Early Apollo Surface Experiment Package (EASEP). Training in lunar surface EVA included practice sessions with lunar surface sample gathering tools and return

containers, cameras, the erectable S-band antenna and the modular equipment stowage assembly (MESA) housed in the LM descent stage.

* Proficiency flights in the lunar landing training vehicle (LLTV) for the commander.

* Zero-g aircraft flights using command module and lunar module mockups for EVA and pressure suit doffing/donning practice and training.

* Underwater zero-g training in the MSC Water Immersion Facility using spacecraft mockups to further familiarize crew with all aspects of CSM-LM docking tunnel intravehicular transfer and EVA in pressurized suits.

* Water egress training conducted in indoor tanks as well as in the Gulf of Mexico included uprighting from the Stable II position (apex down) to the Stable I position (apex up), egress onto rafts and helicopter pickup.

Launch pad egress training from mockups and from the actual spacecraft on the launch pad for possible emergencies such as fire, contaminants and power failures.

* The training covered use of Apollo spacecraft fire suppression equipment in the cockpit.

* Planetarium reviews at Morehead Planetarium, Chapel Hill, N.C., and at Griffith Planetarium, Los Angeles, Calif., of the celestial sphere with special emphasis on the 37 navigational stars used by the Apollo guidance computer.

NATIONAL AERONAUTICS AND SPACE ADMINISTRATION WASHINGTON, D. C. 20546

BIOGRAPHICAL DATA

NAME: Neil A. Armstrong (Mr.)   **NASA Astronaut, Commander, Apollo 11**

BIRTHPLACE AND DATE: Born in Wapakoneta, Ohio, on August 5. 1930; he is the son of Mr. and Mrs. Stephen Armstrong of Wapakoneta.

PHYSICAL DESCRIPTION: Blond hair; blue eyes; height: 5 feet 11 inches; weight: 165 pounds.

EDUCATION: Attended secondary school in Wapakoneta, Ohio; received a Bachelor of Science degree in Aeronautical Engineering from Purdue University in 1955. Graduate School - University of Southern California.

MARITAL STATUS: Married to the former Janet Shearon of Evanston, Illinois, who is the daughter of Mrs. Louise Shearon of Pasadena, California.

CHILDREN: Eric. June 30, 1957; Mark, April 8. 1963.

OTHER ACTIVITIES: His hobbies include soaring (for which he is a Federation Aeronautique Internationale gold badge holder).

ORGANIZATIONS: Associate Fellow of the Society of Experimental Test Pilots; associate fellow of the American Institute of Aeronautics and Astronautics; and member of the Soaring Society of America.

SPECIAL HONORS: Recipient of the 1962 Institute of Aerospace Sciences Octave Chanute Award; the 1966 AIAA Astronautics Award; the NASA Exceptional Service Medal; and the 1962 John J. Montgomery Award.

EXPERIENCE: Armstrong was a naval aviator from 1949 to 1952 and flew 78 combat missions during the Korean action.

He joined NASA's Lewis Research Center in 1955 (then NACA Lewis Flight Propulsion Laboratory) and later transferred to the NASA High Speed Flight Station (now Flight Research Center) at Edwards Air Force Base, California, as an aeronautical research pilot for NACA and NASA. In this capacity, he performed as an X-15 project pilot, flying that aircraft to over 200,000 feet and approximately 4,000 miles per hour.

Other flight test work included piloting the X-1 rocket airplane, the F100, F-101, F-102, F-104, F5D, B-47, the paraglider, and others.

As pilot of the B-29 "drop" aircraft, he participated in the launches of over 100 rocket airplane flights.

He has logged more than 4,000 hours flying time.

CURRENT ASSIGNMENT: Mr. Armstrong was selected as an astronaut by NASA in September 1962. He served as backup command pilot for the Gemini 5 flight.

As command pilot for the Gemini 8 mission, which was launched on March 16, 1966, he performed the first successful docking of two vehicles in space. The flight, originally scheduled to last three days, was terminated early due to a malfunctioning CAMS thruster; but the crew demonstrated exceptional piloting skill in overcoming this problem and bringing the spacecraft to a safe landing.

He subsequently served as backup command pilot for the Gemini 11 mission and is currently assigned as the commander for the Apollo 11 mission, and will probably be the first human to set foot on the Moon.

As a civil servant, Armstrong, a GS-16 Step 7, earns $30,054 per annum

NAME: Michael Collins (Lieutenant Colonel, USAF) NASA Astronaut, Command Module Pilot, Apollo 11

BIRTHPLACE AND DATE: Born in Rome, Italy, on October 31, 1930. His mother, Mrs. James L. Collins, resides in Washington, D.C.

PHYSICAL DESCRIPTION: Brown hair; brown eyes; height: 5 feet 11 inches, weight: 165 pounds.

EDUCATION: Graduated from Saint Albans School in Washington, D.C.; received a Bachelor of Science degree from the United States Military Academy at West Point, New York, in 1952.

MARITAL STATUS: Married to the former Patricia M. Finnegan of Boston, Massachusetts.

CHILDREN: Kathleen, May 6, 1959; Ann S., October 31, 1961; Michael L., February 23, 1963.

OTHER ACTIVITIES: His hobbies include fishing and handball.

ORGANIZATIONS: Member of the Society of Experimental Test Pilots.

SPECIAL HONORS: Awarded the NASA Exceptional Service Medal, the Air Force Command Pilot Astronaut Wings, and the Air Force Distinguished Flying Cross.

EXPERIENCE: Collins, an Air Force Lt. Colonel, chose an Air Force career following graduation from West Point.

He served as an experimental flight test officer at the Air Force Flight Test Center., Edwards Air Force Base, California, and, in that capacity, tested performance and stability and control characteristics of Air Force aircraft— primarily jet fighters. He has logged more than 4,000 hours flying time, including more than 3,200 hours in jet aircraft.

CURRENT ASSIGNMENT: Lt. Colonel Collins was one of the third group of astronauts named by NASA in October 1963. He has since served as backup pilot for the Gemini 7 mission.

As pilot on the 3-day 44-revolution Gemini 10 mission, launched July 18, 1966, Collins shares with command pilot John Young in the accomplishments of that record-setting flight. These accomplishments include a successful rendezvous and docking with a separately launched Agena target vehicle and, using the power of the Agena, maneuvering the Gemini spacecraft into another orbit for a rendezvous with a second, passive Agena. Collins' skillful performance in completing two periods of extravehicular activity, including his recovery of a micrometeorite detection experiment from the passive Agena, added greatly to our knowledge of manned space flight.

Gemini 10 attained an apogee of approximately 475 statute miles and traveled a distance of 1,275,091 statute miles after which splashdown occurred in the West Atlantic 529 statute miles east of Cape Kennedy. The spacecraft landed 2.6 miles from the USS GUADALCANAL and became the second in the Gemini program to land within eye and camera range of a prime recovery vessel.

He is currently assigned as command module pilot on the Apollo 11 mission. The annual pay and allowances of an Air Force lieutenant colonel with Collins' time in service totals $17,147.36.

NAME: Edwin E. Aldrin, Jr. (Colonel., USAF) NASA Astronaut, Lunar Module Pilot, Apollo 11

BIRTHPLACE AND DATE: Born in Montclair, New Jersey, on January 20, 1930, and is the son of the late Marion Moon Aldrin and Colonel (USAF Retired) Edwin E. Aldrin, who resides in Brielle, New Jersey.

PHYSICAL DESCRIPTION: Blond hair; blue eyes; height: 5 feet 10 inches; weight: 165 pounds.

EDUCATION: Graduated from Montclair High School, Montclair, New Jersey; received a Bachelor of Science degree from the United States Military Academy at West Point, New York, in 1951 and a Doctor of Science degree in Astronautics from the Massachusetts Institute of Technology in 1963; recipient of an Honorary Doctorate of Science degree from Gustavus Adolphus College in 1967, Honorary degree from Clark University, Worchester, Mass.

MARITAL STATUS: Married to the former Joan A. Archer of Ho-Ho-Kus, New Jersey, whose parents, Mr. and Mrs. Michael Archer, are residents of that city.

CHILDREN: J. Michael, September 2, 1955; Janice R., August 16, 1957; Andrew J., June 17, 1958.

OTHER ACTIVITIES: He is a Scout Merit Badge Counsellor and an Elder and Trustee of the Webster Presbyterian Church. His hobbies include running, scuba diving, and high bar exercises.

ORGANIZATIONS: Associate Fellow of the American Institute of Aeronautics and Astronautics; member of the Society of Experimental Test Pilots, Sigma Gamma Tau (aeronautical engineering society), Tau Beta Pi (national engineering society), and Sigma Xi (national science research society) and a 32nd Degree Mason advanced through the Commandery and Shrine.

SPECIAL HONORS: Awarded the Distinguished Flying Cross with one Oak Leaf Cluster, the Air Medal with two Oak Leaf Clusters, the Air Force Commendation Medal, the NASA Exceptional Service Medal and Air Force Command Pilot Astronaut Wings, the NASA Group Achievement Award for Rendezvous Operations Planning Team, an Honorary Life Membership in the International Association of Machinists and Aerospace Workers, and an Honorary Membership in the Aerospace Medical Association.

EXPERIENCE: Aldrin, an Air Force Colonel, was graduated third in a class of 475 from the United States Military Academy at West Point in 1951 and subsequently received his wings at Bryan, Texas, in 1952.

He flew 66 combat missions in F-86 aircraft while on duty in Korea with the 51st Fighter Interceptor Wing and was credited with destroying two MIG-15 aircraft. At Nellis Air Force Base, Nevada, he served as an aerial gunnery instructor and then attended the Squadron Officers' School at the Air University, Maxwell Air Force Base, Alabama.

Following his assignment as Aide to the Dean of Faculty at the United States Air Force Academy, Aldrin flew F-100 aircraft as a flight commander with the 36th Tactical Fighter Wing at Bitburg, Germany. He attended MIT, receiving a doctorate after completing his thesis concerning guidance for manned orbital rendezvous, and was then assigned to the Gemini Target Office of the Air Force Space Systems Division, Los Angeles, California. He was later transferred to the USAF Field office at the Manned Spacecraft Center which was responsible for integrating DOD experiments into the NASA Gemini flights.

He has logged approximately 3,500 hours flying time, including 2,853 hours in jet aircraft and 139 hours in helicopters. He has made several flights in the lunar landing research vehicle.

CURRENT ASSIGNMENT: Colonel Aldrin was one of the third group of astronauts named by NASA in October 1963. He has since served as backup pilot for the Gemini 9 mission and prime pilot for the Gemini 12 mission.

On November 11, 1966, he and command pilot James Lovell were launched into space in the Gemini 12 spacecraft on a 4-day 59 revolution flight which brought the Gemini Program to a successful close. Aldrin established a new record for extravehicular activity (EVA) by accruing slightly more than 5½ hours outside the spacecraft. During the umbilical EVA, he attached a tether to the Agena; retrieved a micro-meteorite experiment package from the spacecraft; and evaluated the use of body restraints specially designed for completing work tasks outside the spacecraft. He completed numerous photographic experiments and obtained the first pictures taken from space of an eclipse of the sun.

Other major accomplishments of the 94-hour 35-minute flight included a third-revolution rendezvous with the previously launched Agena, using for the first time backup onboard computations due to a radar failure, and a fully automatic controlled reentry of a spacecraft. Gemini 12 splashed down in the Atlantic within 2½ miles of the prime recovery ship USS WASP.

Aldrin is currently assigned as lunar module pilot for the Apollo 11 flight. The annual pay and allowances of an Air Force colonel with Aldrin's time in service total $18,622-56.

EARLY APOLLO SCIENTIFIC EXPERIMENTS PACKAGE (EASEP)

The Apollo 11 scientific experiments for deployment on the lunar surface near the touchdown point of the lunar module are stowed in the LM's scientific equipment bay at the left rear quadrant of the descent stage looking forward.

The Early Apollo Scientific Experiments Package (EASEP) will be carried only on Apollo 11; subsequent Apollo lunar landing missions will carry the more comprehensive Apollo Lunar Surface Experiment Package (ALSEP).

EASEP consists of two basic experiments: the passive seismic experiments package (PSEP) and the laser ranging retro-reflector (LRRR). Both experiments are independent, self-contained packages that weigh a total of about 170 pounds and occupy 12 cubic feet of space.

PSEP uses three long-period seismometers and one short-period vertical seismometer for measuring meteoroid impacts and moonquakes. Such data will be useful in determining the interior structure of the Moon; for example, does the Moon have a core and mantle like Earth.

The seismic experiment package has four basic subsystems: structure/thermal subsystem for shock, vibration and thermal protection; electrical power subsystem generates 34 to 46 watts by solar panel array; data subsystem receives and decodes MSFN uplink commands and downlinks experiment data, handles power switching tasks; passive seismic experiment subsystem measures lunar seismic activity with long-period and short-period seismometers which detect inertial mass displacement.

The laser ranging retro-reflector experiment is a retroreflector array with a folding support structure for aiming and aligning the array toward Earth. The array is built of cubes of fused silica. Laser ranging beams from Earth will be reflected back to their point of origin for precise measurement of Earth-Moon distances, motion of the Moon's center of mass, lunar radius and Earth geophysical information.

Earth stations which will beam lasers to the LRRR include the McDonald Observatory at Ft. Davis, Tex.; Lick Observatory, Mt. Hamilton, Calif.; and the Catalina Station of the University of Arizona. Scientists in other countries also plan to bounce laser beams off the LRRR.

Principal investigators for these experiments are Dr. C.O. Alley, University of Maryland (Laser Ranging Retro reflector) and Dr. Garry Latham, Lamont Geological Observatory (Passive Seismic Experiments Package).

EASEP/LM INTERFACE

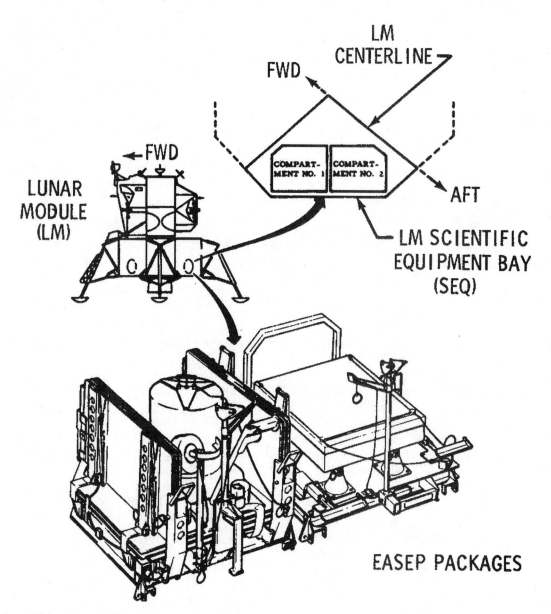

EASEP PACKAGES

EASEP DEPLOYMENT ZONES

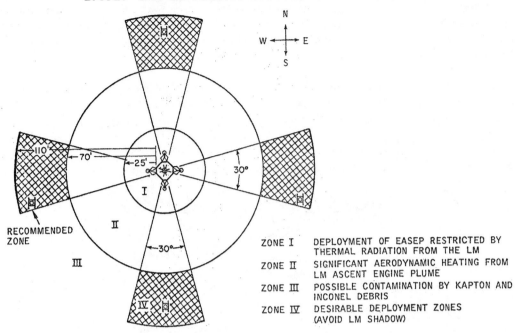

ZONE I   DEPLOYMENT OF EASEP RESTRICTED BY THERMAL RADIATION FROM THE LM
ZONE II   SIGNIFICANT AERODYNAMIC HEATING FROM LM ASCENT ENGINE PLUME
ZONE III POSSIBLE CONTAMINATION BY KAPTON AND INCONEL DEBRIS
ZONE IV DESIRABLE DEPLOYMENT ZONES (AVOID LM SHADOW)

PSEP STOWED CONFIGURATION

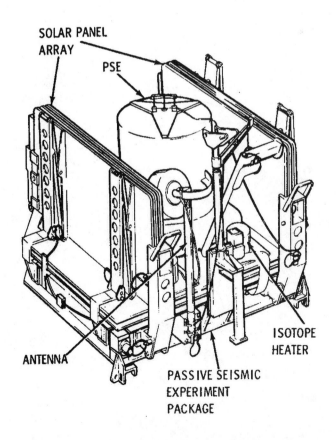

SOLAR PANEL ARRAY

PSE

ANTENNA

PASSIVE SEISMIC EXPERIMENT PACKAGE

ISOTOPE HEATER

PSEP DEPLOYED CONFIGURATION

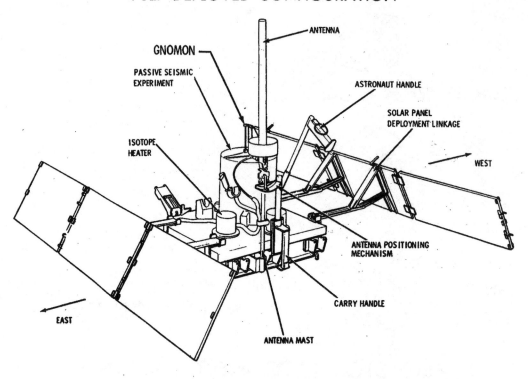

LASER RANGING RETRO-REFLECTOR EXPERIMENT

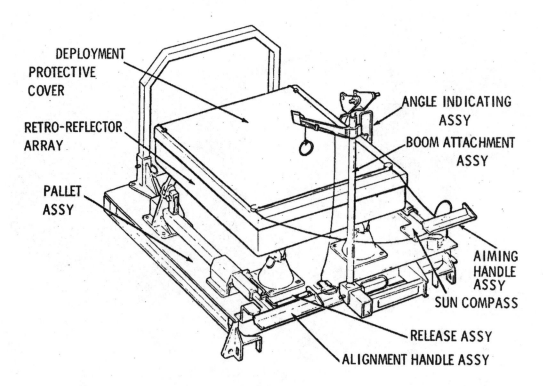

LRRR DETAILS

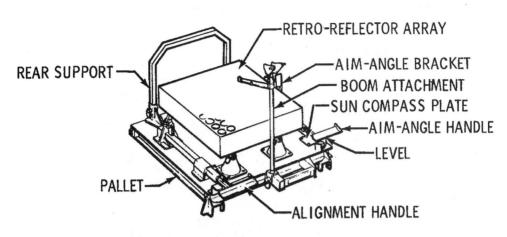

RETRO-REFLECTOR ARRAY

REAR SUPPORT

AIM-ANGLE BRACKET

BOOM ATTACHMENT

SUN COMPASS PLATE

AIM-ANGLE HANDLE

LEVEL

PALLET

ALIGNMENT HANDLE

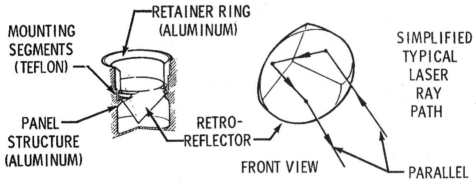

RETAINER RING (ALUMINUM)

MOUNTING SEGMENTS (TEFLON)

PANEL STRUCTURE (ALUMINUM)

RETRO-REFLECTOR

FRONT VIEW

SIMPLIFIED TYPICAL LASER RAY PATH

PARALLEL

SOLAR WIND EXPERIMENT

One of the first rock samples inside a containment vessel at the LRL (above).

COMMAND AND TELEMETRY LINKS

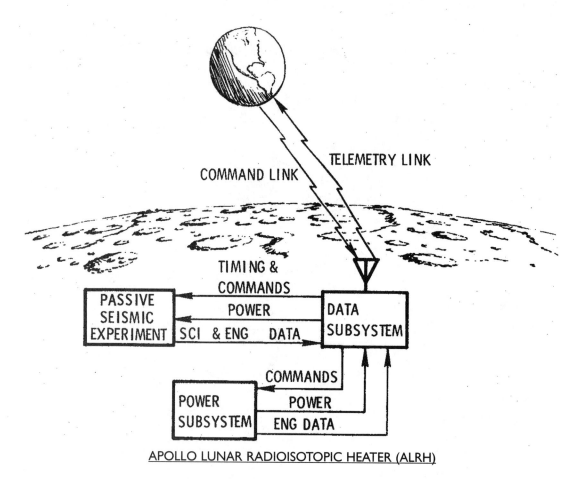

APOLLO LUNAR RADIOISOTOPIC HEATER (ALRH)

An isotopic heater system built into the passive seismometer experiment package which Apollo 11 astronauts will leave on the Moon will protect the seismic recorder during frigid lunar nights.

The Apollo Lunar Radioisotopic Heater (ALRH), developed by the Atomic Energy Commission, will be the first major use of nuclear energy in a manned space flight mission. Each of the two heaters is fueled with about 1.2 ounces of plutonium 238. Heat is given off as the well shielded radioactive material decays.

During the lunar day, the seismic device will send back to Earth data on any lunar seismic activity or "Moonquakes." During the 340-hour lunar night, when temperatures drop as low as 279 degrees below zero F., the 15-watt heaters will keep the seismometer at a minimum of -65 degrees below zero F. Exposure to lower temperatures would damage the device.

Power for the seismic experiment, which operates only during the day, is from two solar panels.

The heaters are three inches in diameter, three inches long, and weigh two pounds and two ounces each including multiple layers of shielding and protective materials. The complete seismometer package weighs 100 pounds.

They are mounted into the seismic package before launch. The entire unit will be carried in the lunar module scientific equipment bay and after landing on the Moon will be deployed by an astronaut a short distance from the lunar vehicle. There is no handling risk to the astronaut.

The plutonium fuel is encased in various materials chosen for radiation shielding and for heat and shock resistance. The materials include a tantalum-tungsten alloy, a platinum-rhodium alloy, titanium, fibrous carbon, and graphite. The outside layer is stainless steel.

Extensive safety analyses and tests were performed by Sandia Laboratories at Albuquerque, New Mexico, to determine effects of an abort or any conceivable accident in connection with the Moon flight. The safety report by the Interagency Safety Evaluation Panel, which is made up of representatives of NASA, the AEC, and the Department of Defense, concluded that the heater presents no undue safety problem to the general population under any accident condition deemed possible for the Apollo mission.

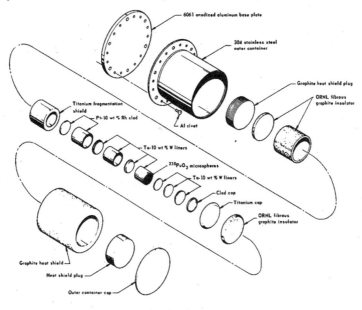

EXPLODED VIEW, APOLLO LUNAR RADIOISOTOPIC HEATER

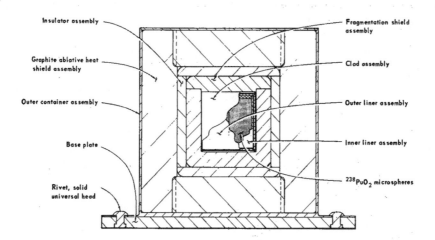

THE APOLLO LUNAR RADIOISOTOPIC HEATER

Sandia Laboratories is operated for the AEC by Western Electric Company. The heater was fabricated by AEC's Mound Laboratory at Miamisburg, Ohio, which is operated by Monsanto Research Corporation.

The first major use of nuclear energy in space came in 1961 with the launching of a navigation satellite with an isotopic generator. Plutonium 238 fuels the device which is still operating. Two similar units were launched in 1961 and two more in 1963.

Last April, NASA launched Nimbus III, a weather satellite with a 2-unit nuclear isotopic system for generating electrical power. The Systems for Nuclear Auxiliary Power (SNAP-19) generator, developed by AEC, provides supplementary power.

Apollo 12 is scheduled to carry a SNAP-27 radioisotope thermoelectric generator, also developed by AEC, to provide power to operate the Apollo Lunar Surface Experiments Package (ALSEP). The SNAP-27 also contains plutonium 238 as the heat source. Thermoelectric elements convert this heat directly into electrical energy.

APOLLO LAUNCH OPERATIONS

Prelaunch Preparations

NASA's John F. Kennedy Space Center performs preflight checkout, test and launch of the Apollo 11 space vehicle. A government-industry team of about 500 will conduct the final countdown from Firing Room 1 of the Launch Control Center (LCC).

The firing room team is backed up by more than 5,000 persons who are directly involved in launch operations at KSC from the time the vehicle and spacecraft stages arrive at the Center until the launch is completed.

Initial checkout of the Apollo spacecraft is conducted in work stands and in the altitude chambers in the Manned Spacecraft Operations Building (MSOB) at Kennedy Space Center. After completion of checkout there, the assembled spacecraft is taken to the Vehicle Assembly Building (VAB) and mated with the launch vehicle. There the first integrated spacecraft and launch vehicle tests are conducted. The assembled space vehicle is then rolled out to the launch pad for final preparations and countdown to launch.

In early January, 1969, flight hardware for Apollo 11 began arriving at Kennedy Space Center, just as Apollo 9 and Apollo 10 were undergoing checkout at KSC.

The lunar module was the first piece of Apollo 11 flight hardware to arrive at KSC. The two stages of the LM were moved into the altitude chamber in the Manned Spacecraft Operations Building after an initial receiving inspection in January. In the chamber the LM underwent systems tests and both unmanned and manned chamber runs. During these runs the chamber air was pumped out to simulate the vacuum of space at altitudes in excess of 200,000 feet. There the spacecraft systems and the astronauts' life support systems were tested.

While the LM was undergoing preparation for its manned altitude chamber runs, the Apollo 11 command/service module arrived at KSC and after receiving inspection, it too was placed in an altitude chamber in the MSOB for systems tests and unmanned and manned chamber runs. The prime and backup crews participated in the chamber runs on both the LM and the CSM.

In early April, the LM and CSM were removed from the chambers. After installing the landing gear on the LM and the SPS engine nozzle on the CSM, the LM was encapsulated in the spacecraft LM adapter (SLA) and the CSM was mated to the SLA. On April 14, the assembled spacecraft was moved to the VAB where it was mated to the launch vehicle.

The launch vehicle flight hardware began arriving at KSC in mid-January and by March 5 the three stages and the instrument unit were erected on Mobile Launcher 1 in high bay 1. Tests were conducted on individual systems on each of the stages and on the overall launch vehicle before the spacecraft was erected atop the vehicle.

After spacecraft erection, the spacecraft and launch vehicle were electrically mated and the first overall test (plugs-in) of the space vehicle was conducted. In accordance with the philosophy of accomplishing as much of the checkout as possible in the VAB, the overall test was conducted before the space vehicle was moved to the launch pad.

The plugs-in test verified the compatibility of the space vehicle systems, ground support equipment and off-site support facilities by demonstrating the ability of the systems to proceed through a simulated countdown, launch and flight. During the simulated flight portion of the test, the systems were required to respond to both emergency and normal flight conditions.

The move to Pad A from the VAB on May 21 occurred while Apollo 10 was enroute to the Moon for a dress rehearsal of a lunar landing mission and the first test of a complete spacecraft in the near-lunar environment.

Apollo 11 will mark the fifth launch at Pad A on Complex 39. The first two unmanned Saturn V launches and the manned Apollo 8 and 9 launches took place at Pad A. Apollo 10 was the only launch to date from Pad B.

The space vehicle Flight Readiness Test was conducted June 4-6. Both the prime and backup crews participate in portions of the FRT, which is a final overall test of the space vehicle systems and ground support equipment when all systems are as near as possible to a launch configuration.

After hypergolic fuels were loaded aboard the space vehicle and the launch vehicle first stage fuel (RP-1) was brought aboard, the final major test of the space vehicle began. This was the countdown demonstration test (CDDT), a dress rehearsal for the final countdown to launch. The CDDT for Apollo 11 was divided into a "wet" and a "dry" portion. During the first, or "wet" portion, the entire countdown, including propellant loading, was carried out down to T-8.9 seconds, the time for ignition sequence start. The astronaut crew did not participate in the wet CDDT.

At the completion of the wet CDDT, the cryogenic propellants (liquid oxygen and liquid hydrogen) were off-loaded, and the final portion of the countdown was re-run, this time simulating the fueling and with the prime astronaut crew participating as they will on launch day.

By the time Apollo 11 was entering the final phase of its checkout procedure at Complex 39A, crews had already started the checkout of Apollo 12 and Apollo 13. The Apollo 12 spacecraft completed altitude chamber testing in June and was later mated to the launch vehicle in the VAB. Apollo 13 flight hardware began arriving in June to undergo preliminary checkout.

Because of the complexity involved in the checkout of the 363-foot-tall (110.6 meters) Apollo/Saturn V configuration, the launch teams make use of extensive automation in their checkout. Automation is one of the major differences in checkout used in Apollo compared to the procedures used in the Mercury and Gemini programs.

Computers, data display equipment and digital data techniques are used throughout the automatic checkout from the time the launch vehicle is erected in the VAB through liftoff. A similar, but separate computer operation called ACE (Acceptance Checkout-Equipment) is used to verify the flight readiness of the spacecraft. Spacecraft checkout is controlled from separate rooms in the Manned Spacecraft Operations Building.

LAUNCH COMPLEX 39

Launch Complex 39 facilities at the Kennedy Space Center were planned and built specifically for the Apollo Saturn V, the space vehicle that will be used to carry astronauts to the Moon.

Complex 39 introduced the mobile concept of launch operations, a departure from the fixed launch pad techniques used previously at Cape Kennedy and other launch sites. Since the early 1950's when the first ballistic missiles were launched, the fixed launch concept had been used on NASA missions. This method called for assembly, checkout and launch of a rocket at one site— the launch pad. In addition to tying up the pad, this method also often left the flight equipment exposed to the outside influences of the weather for extended periods.

Using the mobile concept, the space vehicle is thoroughly checked in an enclosed building before it is moved to the launch pad for final preparations. This affords greater protection, a more systematic checkout process using computer techniques and a high launch rate for the future, since the pad time is minimal.

Saturn V stages are shipped to the Kennedy Space Center by ocean-going vessels and specially designed aircraft, such as the Guppy. Apollo spacecraft modules are transported by air. The spacecraft components are first taken to the Manned Spacecraft Operations Building for preliminary checkout. The Saturn V stages are brought immediately to the Vehicle Assembly Building after arrival at the nearby turning basin.

The major components of Complex 39 include: (1) the Vehicle Assembly Building (VAB) where the Apollo 11 was assembled and prepared; (2) the Launch Control Center, where the launch team conducts the preliminary checkout and final countdown; (3) the mobile launcher, upon which the Apollo 11 was erected for checkout and from where it will be launched; (4) the mobile service structure, which provides external access to the space vehicle at the pad; (5) the transporter, which carries the space vehicle and mobile launcher, as well as the mobile service structure to the pad; (6) the crawlerway over which the space vehicle travels from the VAB to the launch pad; and (7) the launch pad itself.

Vehicle Assembly Building

The Vehicle Assembly Building is the heart of Launch Complex 39. Covering eight acres, it is where the 363-foot-tall space vehicle is assembled and tested.

The VAB contains 129,482,000 cubic feet of space. It is 716 feet long, and 518 feet wide and it covers 343,500 square feet of floor space.

The foundation of the VAB rests on 4,225 steel pilings, each 16 inches in diameter, driven from 150 to 170 feet to bedrock. If placed end to end these pilings would extend a distance of 123 miles. The skeletal structure of the building contains approximately 60,000 tons of structural steel. The exterior is covered by more than a million square feet of insulated aluminum siding.

The building is divided into a high bay area 525 feet high and a low bay area 210 feet high, with both areas serviced by a transfer aisle for movement of vehicle stages.

The low bay work area, approximately 442 feet wide and 274 feet long, contains eight stage-preparation and checkout cells. These cells are equipped with systems to simulate stage interface and operation with other stages and the instrument unit of the Saturn V launch vehicle.

After the Apollo 11 launch vehicle upper stages arrived at Kennedy Space Center, they were moved to the low bay of the VAB. Here, the second and third stages underwent acceptance and checkout testing prior to mating with the S-IC first stage atop the Mobile Launcher in the high bay area.

The high bay provides facilities for assembly and checkout of both the launch vehicle and spacecraft. It contains four separate bays for vertical assembly and checkout. At present, three bays are equipped, and the fourth will be reserved for possible changes in vehicle configuration.

Work platforms, some as high as three-story buildings, in the high bays provide access by surrounding the vehicle at varying levels. Each high bay has five platforms. Each platform consists of two bi-parting sections that move in from opposite sides and mate, providing a 360 degree access to the section of the space vehicle being checked.

A 10,000-ton-capacity air conditioning system, sufficient to cool about 3,000 homes, helps to control the environment within the entire office, laboratory, and workshop complex located inside the low bay area of the VAB. Air conditioning is also fed to individual platform levels located around the vehicle.

There are 141 lifting devices in the VAB, ranging from one-ton hoists to two 250-ton high-lift bridge cranes.

The mobile launchers, carried by transporter vehicles, move in and out of the VAB through four doors in the high bay area, one in each of the bays. Each door is shaped like an inverted T. They are 152 feet wide and 114 feet high at the base, narrowing to 76 feet in width. Total door height is 456 feet.

The lower section of each door is of the aircraft hangar type that slides horizontally on tracks. Above this are seven telescoping vertical lift panels stacked one above the other, each 50 feet high and driven by an individual motor. Each panel slides over the next to create an opening large enough to permit passage of the mobile launcher.

Launch Control Center

Adjacent to the VAB is the Launch Control Center (LCC). This four-story structure is a radical departure from the dome-shaped blockhouses at other launch sites.

The electronic "brain" of Launch Complex 39, the LCC was used for checkout and test operations while Apollo 11 was being assembled inside the VAB. The LCC contains display, monitoring, and control equipment used for both checkout and launch operations.

The building has telemeter checkout stations on its second floor, and four firing rooms, one for each high bay of the VAB on its third floor. Three firing rooms contain identical sets of control and monitoring equipment, so that launch of a vehicle and checkout of others take place simultaneously. A ground computer facility is associated with each firing room.

The high speed computer data link is provided between the LCC and the mobile launcher for checkout of the launch vehicle. This link can be connected to the mobile launcher at either the VAB or at the pad.

The three equipped firing rooms have some 450 consoles which contain controls and displays required for the checkout process. The digital data links connecting with the high bay areas of the VAB and the launch pads carry vast amounts of data required during checkout and launch.

There are 15 display systems in each LCC firing room, with each system capable of providing digital information instantaneously.

Sixty television cameras are positioned around the Apollo/ Saturn V transmitting pictures on 10 modulated channels. The LCC firing room also contains 112 operational intercommunication channels used by the crews in the checkout and launch countdown.

Mobile Launcher

The mobile launcher is a transportable launch base and umbilical tower for the space vehicle. Three mobile launchers are used at Complex 39.

The launcher base is a two-story steel structure, 25 feet high, 160 feet long, and 135 feet wide. It is positioned on six steel pedestals 22 feet high when in the VAB or at the launch pad. At the launch pad, in addition to the six steel pedestals, four extendable columns also are used to stiffen the mobile launcher against rebound loads, if the Saturn engines cut off.

The umbilical tower, extending 398 feet above the launch platform, is mounted on one end of the launcher base. A hammerhead crane at the top has a hook height of 376 feet above the deck with a traverse radius of 85 feet from the center of the tower.

The 12-million-pound mobile launcher stands 445 feet high when resting on its pedestals. The base, covering about half an acre, is a compartmented structure built of 25-foot steel girders.

The launch vehicle sits over a 45-foot-square opening which allows an outlet for engine exhausts into the launch pad trench containing a flame deflector. This opening is lined with a replaceable steel blast shield, independent of the structure, and is cooled by a water curtain initiated two seconds after liftoff.

There are nine hydraulically-operated service arms on the umbilical tower. These service arms support lines for the vehicle umbilical systems and provide access for personnel to the stages as well as the astronaut crew to the spacecraft.

On Apollo 11, one of the service arms is retracted early in the count. The Apollo spacecraft access arm is partially retracted at T-43 minutes. A third service arm is released at T-30 seconds, and a fourth at about T-16.5 seconds. The remaining five arms are set to swing back at vehicle first motion after T-0.

The service arms are equipped with a backup retraction system in case the primary mode fails.

The Apollo access arm (service arm 9), located at the 320 foot level above the launcher base, provides access to the spacecraft cabin for the closeout team and astronaut crews. The flight crew will board the spacecraft starting about T-2 hours, 40 minutes in the count. The access arm will be moved to a parked position, 12 degrees from the spacecraft, at about T-43 minutes. This is a distance of about three feet, which permits a rapid reconnection of the arm to the spacecraft in the event of an emergency condition. The arm is fully retracted at the T-5 minute mark in the count.

The Apollo 11 vehicle is secured to the mobile launcher by four combination support and hold-down arms mounted on the launcher deck. The hold-down arms are cast in one piece, about 6 x 9 feet at the base and 10 feet tall, weighing more than 20 tons. Damper struts secure the vehicle near its top.

After the engines ignite, the arms hold Apollo 11 for about six seconds until the engines build up to 95 percent thrust and other monitored systems indicate they are functioning properly. The arms release on receipt of a launch commit signal at the zero mark in the count. But the vehicle is prevented from accelerating too rapidly by controlled release mechanisms.

The mobile launcher provides emergency egress for the crew and closeout service personnel. Personnel may descend the tower via two 600-feet per minute elevators or by a slide-wire and cab to a bunker 2,200 feet from the launcher. If high speed elevators are utilized to level A of the launcher, two options are then available. The personnel may slide down the escape tube to the blast room below the pad or take elevator B to the bottom of the pad and board armored personnel carriers and depart the area.

Transporter

The six-million-pound transporters move mobile launchers into the VAB and mobile launchers with assembled Apollo space vehicles to the launch pad. They also are used to transfer the mobile service structure to and from the launch pads. Two transporters are in use at complex 39.

The transporter is 131 feet long and 114 feet wide. The vehicle moves on four doubletracked crawlers, each 10 feet high and 40 feet long. Each shoe on the crawler track is seven feet six inches in length and weighs about a ton.

Sixteen traction motors powered by four 1,000-kilowatt generators, which in turn are driven by two 2,750-horsepower diesel engines, provide the motive power for the transporter. Two 750kw generators, driven by two 1,065-horsepower diesel engines, power the jacking, steering, lighting, ventilating and electronic systems.

Maximum speed of the transporter is about one-mile-per-hour loaded and about two-miles-per-hour unloaded. The 3.5 mile trip to Pad A with Apollo 11 on its mobile launcher took about six hours since maximum speed is not maintained throughout the trip.

The transporter has a leveling system designed to keep the top of the space vehicle vertical within plus-or-minus 10 minutes of arc — about the dimensions of a basketball.

This system also provides leveling operations required to negotiate the five percent ramp which leads to the launch pad and keeps the load level when it is raised and lowered on pedestals both at the pad and within the VAB.

The overall height of the transporter is 20 feet from ground level to the top deck on which the mobile launcher is mated for transportation. The deck is flat and about the size of a baseball diamond (90 by 90 feet).

Two operator control cabs, one at each end of the chassis located diagonally opposite each other, provide totally enclosed stations from which all operating and control functions are coordinated.

Crawlerway

The transporter moves on a roadway 131 feet wide, divided by a median strip. This is almost as broad as an eight-lane turnpike and is designed to accommodate a combined weight of about 18 million pounds.

The roadway is built in three layers with an average depth of seven feet. The roadway base layer is two-and-one-half feet of hydraulic fill compacted to 95 percent density. The next layer consists of three feet of crushed rock packed to maximum density, followed by a layer of one foot of selected hydraulic fill. The bed is topped and sealed with an asphalt prime coat.

On top of the three layers is a cover of river rock, eight inches deep on the curves and six inches deep on the straightway. This layer reduces the friction during steering and helps distribute the load on the transporter bearings.

Mobile Service Structure

A 402-foot-tall, 9.8-million-pound tower is used to service the Apollo launch vehicle and spacecraft at the pad. The 40-story steel-trussed tower, called a mobile service structure, provides 360-degree platform access to the Saturn launch vehicle and the Apollo spacecraft.

The service structure has five platforms — two self-propelled and three fixed, but movable. Two elevators carry personnel and equipment between work platforms. The platforms can open and close around the 363-foot space vehicle.

After depositing the mobile launcher with its space vehicle on the pad, the transporter returns to a parking area about 7,000 feet from pad A. There it picks up the mobile service structure and moves it to the launch pad. At the pad, the huge tower is lowered and secured to four mount mechanisms.

The top three work platforms are located in fixed positions which serve the Apollo spacecraft. The two lower movable platforms serve the Saturn V.

The mobile service structure remains in position until about T-11 hours when it is removed from its mounts and returned to the parking area.

Water Deluge System

A water deluge system will provide a million gallons of industrial water for cooling and fire prevention during launch of Apollo 11. Once the service arms are retracted at liftoff, a spray system will come on to cool these arms from the heat of the five Saturn F-1 engines during liftoff.

On the deck of the mobile launcher are 29 water nozzles. This deck deluge will start immediately after liftoff and will pour across the face of the launcher for 30 seconds at the rate of 50,000 gallons-per-minute. After 30 seconds, the flow will be reduced to 20,000 gallons-per-minute.

Positioned on both sides of the flame trench are a series of nozzles which will begin pouring water at 8,000 gallons-per-minute, 10 seconds before liftoff. This water will be directed over the flame deflector.

Other flush mounted nozzles, positioned around the pad, will wash away any fluid spill as a protection against fire hazards.

Water spray systems also are available along the egress route that the astronauts and closeout crews would follow in case an emergency evacuation was required.

Flame Trench and Deflector

The flame trench is 58 feet wide and approximately six feet above mean sea level at the base. The height of the trench and deflector is approximately 42 feet.

The flame deflector weighs about 1.3 million pounds and is stored outside the flame trench on rails. When it is moved beneath the launcher, it is raised hydraulically into position. The deflector is covered with a four-and-one-half-inch thickness of refractory concrete consisting of a volcanic ash aggregate and a calcium aluminate binder. The heat and blast of the engines are expected to wear about three-quarters of an inch from this refractory surface during the Apollo 11 launch.

Pad Areas

Both Pad A and Pad B of Launch complex 39 are roughly octagonal in shape and cover about one fourth of a square mile of terrain.

The center of the pad is a hardstand constructed of heavily reinforced concrete. In addition to supporting the weight of the mobile launcher and the Apollo Saturn V vehicle, it also must support the 9.8-million-pound mobile service structure and 6 million-pound transporter, all at the same time. The top of the pad stands some 48 feet above sea level.

Saturn V propellants — liquid oxygen, liquid hydrogen and RP-1 — are stored near the pad perimeter.

Stainless steel, vacuum-jacketed pipes carry the liquid oxygen (LOX) and liquid hydrogen from the storage tanks to the pad, up the mobile launcher, and finally into the launch vehicle propellant tanks.

LOX is supplied from a 900,000-gallon storage tank. A centrifugal pump with a discharge pressure of 320 pounds-per-square-inch pumps LOX to the vehicle at flow rates as high as 10,000-gallons-per-minute.

Liquid hydrogen, used in the second and third stages, is stored in an 850,000-gallon tank, and is sent through 1,500 feet of 10-inch, vacuum-jacketed invar pipe. A vaporizing heat exchanger pressurizes the storage tank to 60 psi for a 10,000 gallons-per-minute flow rate.

The RP-1 fuel, a high grade of kerosene is stored in three tanks— each with a capacity of 86,000 gallons. It is pumped at a rate of 2,000 gallons-per-minute at 175 psig.

The Complex 39 pneumatic system includes a converter-compressor facility, a pad high-pressure gas storage battery, a high-pressure storage battery in the VAB, low and high-pressure, cross-country supply lines, high-pressure hydrogen storage and conversion equipment, and pad distribution piping to pneumatic control panels. The various purging systems require 187,000 pounds of liquid nitrogen and 21,000 gallons of helium.

Mission Control Center

The Mission Control Center at the Manned Spacecraft Center, Houston, is the focal point for Apollo flight control activities. The center receives tracking and telemetry data from the Manned Space Flight Network, processes this data through the Mission Control Center Real Time Computer Complex, and displays this data to the flight controllers and engineers in the Mission Operations Control Room and staff support rooms.

The Manned Space Flight Network tracking and data acquisition stations link the flight controllers at the center to the spacecraft.

For Apollo 11 all network stations will be remote sites, that is, without flight control teams. All uplink commands and voice communications will originate from Houston, and telemetry data will be sent back to Houston at high speed rates (2,400 bits-per-second), on two separate data lines. They can be either real time or playback information.

Signal flow for voice circuits between Houston and the remote sites is via commercial carrier, usually satellite, wherever possible using leased lines which are part of the NASA Communications Network.

Commands are sent from Houston to NASA's Goddard Space Flight Center, Greenbelt, Md., on lines which link computers at the two points. The Goddard communication computers provide automatic switching facilities and speed buffering for the command data. Data are transferred from Goddard to remote sites on high speed (2,400 bits-per-second) lines. Command loads also can be sent by teletype from Houston to the remote sites at 100 words-per-minute. Again, Goddard computers provide storage and switching functions.

Telemetry data at the remote site are received by the RF receivers, processed by the pulse code modulation ground stations, and transferred to the 642B remote-site telemetry computer for storage. Depending on the format selected by the telemetry controller at Houston, the 642B will send the desired format through a 2010 data transmission unit which provides parallel to serial conversion, and drives a 2,400 bit-per-second mode.

The data mode converts the digital serial data to phase-shifted keyed tones which are fed to the high speed data lines of the communications network.

Tracking data are sent from the sites in a low speed (100 words) teletype format and a 240-bit block high speed (2,400 bits) format. Data rates are one sample-6 seconds for teletype and 10 samples (frames) per second for high speed data.

All high-speed data, whether tracking or telemetry, which originate at a remote site are sent to Goddard on highspeed lines. Goddard reformats the data when necessary and sends them to Houston in 600-bit blocks at a 40,800 bits-per-second rate. Of the 600-bit block, 480 bits are reserved for data, the other 120 bits for address, sync, intercomputer instructions, and polynominal error encoding.

All wideband 40,800 bits-per-second data originating at Houston are converted to high speed (2,400 bits-per-second) data at Goddard before being transferred to the designated remote site.

MANNED SPACEFLIGHT NETWORK

Tracking, command and communication — Apollo 11's vital links with the Earth — will be performed, in two broad phases.

For the first phase, the Manned Space Flight Network (MSFN) will depend largely on its worldwide chain of stations equipped with 30-foot antennas while Apollo is launched and orbiting near the Earth. The second phase begins when the spacecraft moves out more than 10,000 miles above Earth, when the 85-foot diameter antennas bring their greater power and accuracy into play.

The Network must furnish reliable, instantaneous contact with the astronauts, their launch vehicle and spacecraft, from liftoff through Earth orbit, Moon landing and lunar takeoff to splashdown in the Pacific Ocean.

For Apollo 11, MSFN will use 17 ground stations, four ships and six to eight jet aircraft — all directly or indirectly linked with Mission Control Center in Houston. While the Earth turns on its axis and the Moon travels in orbit nearly one quarter million miles away and Apollo 11 moves between them, ground controllers will be kept in the closest possible contact. Thus, only for some 45 minutes as the spacecraft flies behind the Moon in each orbit, will this link with Earth be out of reach.

All elements of the Network get ready early in the countdown. As the Apollo Saturn V ascends, voice and data will be transmitted instantaneously to Houston. The data are sent directly through computers for visual display to flight controllers.

Depending on the launch azimuth, the 30-foot antennas will keep tabs on Apollo 11, beginning with the station at Merritt Island, thence Grand Bahama Island; Bermuda; tracking ship Vanguard; the Canary Islands; Carnarvon, Australia; Hawaii; another tracking ship; Guaymas, Mexico; and Corpus Christi, Tex.

To inject Apollo 11 into translunar flight path, Mission Control will send a signal through one of the land stations or one of the tracking ships in the Pacific. As the spacecraft heads for the Moon, the engine burn will be monitored by the ships and an Apollo range instrumentation aircraft (ARIA). The ARIA provides a relay for the astronauts' voices and data communication with Houston.

When the spacecraft reaches an altitude of 10,000 miles the more powerful 85-foot antennas will join in for primary support of the flight, although the 30-foot "dishes" will continue to track and record data. The 85-foot antennas are located, about 120 degrees apart, near Madrid, Spain; Goldstone, Calif.; and Canberra, Australia.

With the 120-degree spacing around the Earth, at least one of the large antennas will have the Moon in view at all times. As the Earth revolves from west to east, one 85-foot station hands over control to the next 85-foot station as it moves into view of the spacecraft. In this way, data and communication flow is maintained.

Data are relayed back through the huge antennas and transmitted via the NASA Communications Network (NASCOM) — a two-million mile hookup of landlines, undersea cables, radio circuits and communication satellites to Houston. This information is fed into computers for visual display in Mission Control — for example, a display of the precise position of the spacecraft on a large map. Or, returning data may indicate a drop in power or some other difficulty in a spacecraft system, which would energize a red light to alert a flight controller to action.

Returning data flowing through the Earth stations give the necessary information for commanding midcourse maneuvers to keep Apollo 11 in a proper trajectory for orbiting the Moon. After Apollo 11 is in the vicinity of the Moon, these data indicate the amount of retro burn necessary for the service module engine to place the spacecraft in lunar orbit.

Once the lunar module separates from the command module and goes into a separate lunar orbit, the MSFN

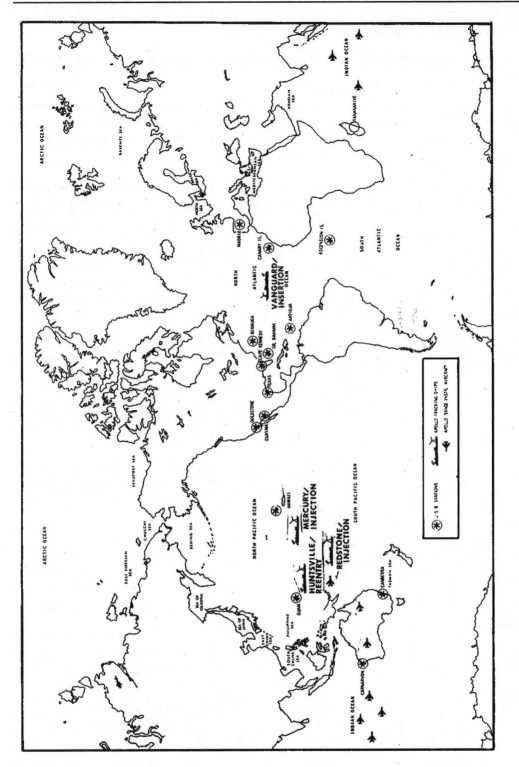

MANNED SPACE FLIGHT TRACKING NETWORK

will be required to keep track of both spacecraft at once, and provide two-way communication and telemetry between them and the Earth. The prime antenna at each of the three 85-foot tracking stations will handle one spacecraft while a wing, or backup, antenna at the same site will handle the other spacecraft during each pass.

Tracking and acquisition of data between Earth and the two spacecraft will provide support for the rendezvous and docking maneuvers. The information will also be used to determine the time and duration of the service module propulsion engine burn required to place the command module into a precise trajectory for reentering the Earth's atmosphere at the planned location.

As the spacecraft comes toward Earth at high speed — up to more than 25,000 miles per hour — it must reenter at the proper angle. To make an accurate reentry, information from the tracking stations and ships is fed into the MCC computers where flight controllers make decisions that will provide the Apollo 11 crew with the necessary information.

Appropriate MSFN stations, including the ships and aircraft in the Pacific, are on hand to provide support during the reentry. An ARIA aircraft will relay astronaut voice communications to MCC and antennas on reentry ships will follow the spacecraft.

Through the journey to the Moon and return, television will be received from the spacecraft at the three 85-foot antennas around the world. In addition, 210-foot diameter antennas in California and Australia will be used to augment the television coverage while the Apollo 11 is near and on the Moon. Scan converters at the stations permit immediate transmission of commercial quality TV via NASCOM to Houston, where it will be released to TV networks.

NASA Communications Network

The NASA Communications Network (NASCOM) consists of several systems of diversely routed communications channels leased on communications satellites, common carrier systems and high frequency radio facilities where necessary to provide the access links.

The system consists of both narrow and wide-band channels, and some TV channels. Included are a variety of telegraph, voice, and data systems (digital and analog) with several digital data rates. Wide-band systems do not extend overseas. Alternate routes or redundancy provide added reliability.

A primary switching center and intermediate switching and control points provide centralized facility and technical control, and switching operations under direct NASA control. The primary switching center is at the Goddard Space Flight Center, Greenbelt, Md. Intermediate switching centers are located at Canberra, Madrid, London, Honolulu, Guam, and Kennedy Space Center.

For Apollo 11, the Kennedy Space Center is connected directly to the Mission Control Center, Houston via the Apollo Launch Data System and to the Marshall Space Flight Center, Huntsville, Ala., by a Launch Information Exchange Facility.

After launch, all network tracking and telemetry data hubs at GSFC for transmission to MCC Houston via two 50,000 bits-per-second circuits used for redundancy and in case of data overflow.

Two Intelsat communications satellites will be used for Apollo 11. The Atlantic satellite will service the Ascension Island unified S-band (USB) station, the Atlantic Ocean ship and the Canary Islands site.

The second Apollo Intelsat communications satellite over the mid-Pacific will service the Carnarvon, Australia USB site and the Pacific Ocean ships. All these stations will be able to transmit simultaneously through the satellite to Houston via Brewster Flat, Wash., and the Goddard Space Flight Center, Greenbelt. Md.

Network Computers

At fraction-of-a-second intervals, the network's digital data processing systems, with NASA's Manned Spacecraft Center as the focal point, "talk" to each other or to the spacecraft. Highspeed computers at the remote site (tracking ships included) issue commands or "up-link" data on such matters as control of cabin pressure, orbital guidance commands, or "go-no-go" indications to perform certain functions.

When information originates from Houston, the computers refer to their preprogrammed information for

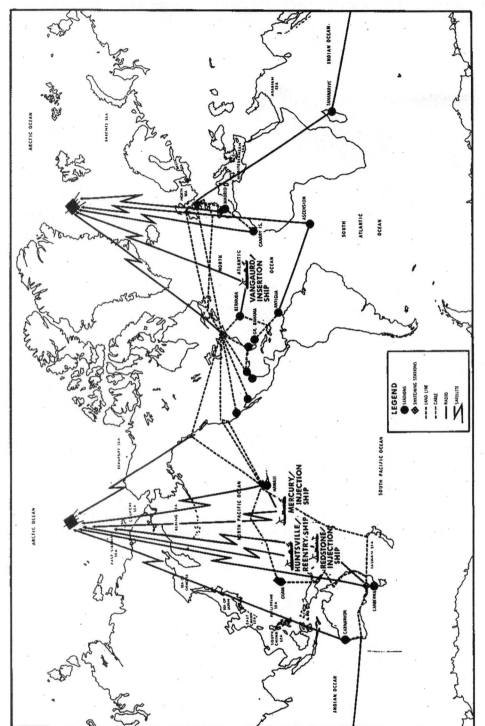

NASA COMMUNICATIONS NETWORK

validity before transmitting the required data to the spacecraft.

Such "up-link" information is communicated by ultra-high-frequency radio about 1,200 bits-per-second. Communication between remote ground sites, via high-speed communications links, occurs at about the same rate. Houston reads information from these ground sites at 2,400 bits-per-second, as well as from remote sites at 100 words-per-minute.

The computer systems perform many other functions, including:

Assuring the quality of the transmission lines by continually exercising data paths.

Verifying accuracy of the messages by repetitive operations.

Constantly updating the flight status.

For "down link" data, sensors built into the spacecraft continually sample cabin temperature, pressure, physical information on the astronauts such as heartbeat and respiration, among other items. These data are transmitted to the ground stations at 51.2 kilobits (12,800 binary digits) per-second.

At MCC the computers:

* Detect and select changes or deviations, compare with their stored programs, and indicate the problem areas or pertinent data to the flight controllers.

* Provide displays to mission personnel.

* Assemble output data in proper formats.

* Log data on magnetic tape for replay for the flight controllers.

* Keep time.

The Apollo Ships

The mission will be supported by four Apollo instrumentation ships operating as integral stations of the Manned Space Flight Network (MSFN) to provide coverage in areas beyond the range of land stations.

The ships, USNS Vanguard, Redstone, Mercury, and Huntsville will perform tracking, telemetry, and communication functions for the launch phase, Earth orbit insertion, translunar injection, and reentry.

Vanguard will be stationed about 1,000 miles southeast of Bermuda (25 degrees N., 49 degrees W.) to bridge the Bermuda-Antigua gap during Earth orbit insertion. Vanguard also functions as part of the Atlantic recovery fleet in the event of a launch phase contingency. Redstone (2.25 degrees S., 166.8 degrees E.); the Mercury (10 N., 175.2 W.) and the Huntsville (3.0 N., 154.0 E.) provide a triangle of mobile stations between the MSFN stations at Carnarvon and Hawaii for coverage of the burn interval for translunar injection. In the event the launch date slips from July 16, the ships will all move generally north-eastward to cover the changing translunar injection location.

Redstone and Huntsville will be repositioned along the reentry corridor for tracking, telemetry, and communications functions during reentry and landing. They will track Apollo from about 1,000 miles away through communications blackout when the spacecraft will drop below the horizon and will be picked up by the ARIA aircraft.

The Apollo ships were developed jointly by NASA and the Department of Defense. The DOD operates the ships in support of Apollo and other NASA and DOD missions on a non-interference basis with Apollo requirements.

Management of the Apollo ships is the responsibility of the Commander, Air Force Western Test Range (AFWTR). The Military Sea Transport Service provides the maritime crews and the Federal Electric Corp., International Telephone and Telegraph, under contract to AFWTR, provides the technical instrumentation crews.

The technical crews operate in accordance with joint NASA/ DOD standards and specifications which are compatible with MSFN operational procedures.

Apollo Range Instrumentation Aircraft (ARIA)

During Apollo 11, the ARIA will be used primarily to fill coverage gaps between the land and ship stations in the Pacific between Australia and Hawaii during the translunar injection interval. Prior to and during the burn, the ARIA record telemetry data from Apollo and provide realtime voice communication between the astronauts and the Mission Control Center at Houston.

Eight aircraft will participate in this mission, operating from Pacific, Australian and Indian Ocean air fields in positions under the orbital track of the spacecraft and launch vehicle. The aircraft will be deployed in a northwestward direction in the event of launch day slips.

For reentry, two ARIA will be deployed to the landing area to continue communications between Apollo and Mission Control at Houston and provide position information on the spacecraft after the blackout phase of reentry has passed.

The total ARIA fleet for Apollo missions consists of eight EC-135A (Boeing 707) jets equipped specifically to meet mission needs. Seven-foot parabolic antennas have been installed in the nose section of the planes giving them a large, bulbous look.

The aircraft, as well as flight and instrumentation crews, are provided by the Air Force and they are equipped through joint Air Force-NASA contract action to operate in accordance with MSFN procedures.

Ship Positions for Apollo 11

July 16, 1969

Insertion Ship (VAN)	25 degrees N 49 degrees W
Injection Ship (MER)	10 degrees N 175.2 degrees W
Injection Ship (RED)	2.25 degrees S 166.8 degrees E
Injection Ship (HTV)	3.0 degrees N 154.0 degrees E
Reentry Support	
Reentry Ship (HTV)	5.5 degrees N 178.2 degrees W
Reentry Ship (RED)	3.0 degrees S 165.5 degrees E

July 18, 1969

Insertion Ship (VAN)	25 degrees N 49 degrees W
Injection Ship (MER)	15 degrees N 166.5 degrees W
Injection Ship (RED)	4.0 degrees N 172.0 degrees E
Injection Ship (HTV)	10.0 degrees N 157.0 degrees E
Reentry Support	
Reentry Ship (HTV)	17.0 degrees N 177.3 degrees W
Reentry Ship (RED)	6.5 degrees N 163.0 degrees E

July 21, 1969

Insertion Ship (VAN)	25 degrees N 49 degrees W
Injection Ship (MER)	16.5 degrees N 151 degrees W
Injection Ship (RED)	11.5 degrees N 177.5 degrees W
Injection Ship (HTV)	12.0 degrees N 166.0 degrees E
Reentry Support	
Reentry Ship (HTV)	26.0 degrees N 176.8 degrees W
Reentry Ship (RED)	17.3 degrees N 160.0 degrees E

CONTAMINATION CONTROL PROGRAM

In 1966 an Interagency Committee on Back Contamination (ICBC) was established. The function of this Committee was to assist NASA in developing a program to prevent the contamination of the Earth from lunar materials following manned lunar exploration. The committee charter included specific authority to review and approve the plans and procedures to prevent back contamination. The committee membership includes representatives from the Public Health Service, Department of Agriculture, Department of the Interior, NASA, and the National Academy of Sciences.

Over the last several years NASA has developed facilities, equipment and operational procedures to provide an adequate back contamination program for the Apollo missions. This program of facilities and procedures, which is well beyond the current state-of-the-art, and the overall effort have resulted in a laboratory with capabilities which have never previously existed. The scheme of isolation of the Apollo crewmen and lunar samples, and the exhaustive test programs to be conducted are extensive in scope and complexity.

The Apollo Back Contamination Program can be divided into three phases. The first phase covers the procedures which are followed by the crew while in flight to reduce and, if possible, eliminate the return of lunar surface contaminants in the command module.

The second phase includes spacecraft and crew recovery and the provisions for isolation and transport of the crew, spacecraft, and lunar samples to the Manned Spacecraft Center. The third phase encompasses the quarantine operations and preliminary sample analysis in the Lunar Receiving Laboratory.

A primary step in preventing back contamination is careful attention to spacecraft cleanliness following lunar surface operations. This includes use of special cleaning equipment, stowage provisions for lunar-exposed equipment, and crew procedures for proper "housekeeping."

Lunar Module Operations - The lunar module has been designed with a bacterial filter system to prevent contamination of the lunar surface when the cabin atmosphere is released at the start of the lunar exploration.

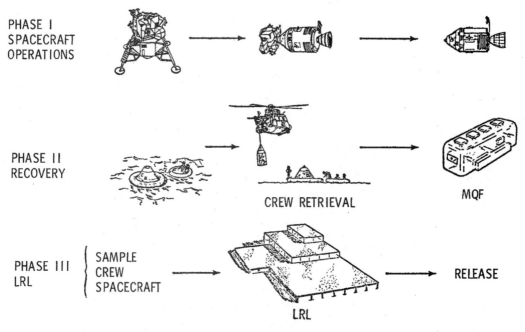

PHASE I
SPACECRAFT
OPERATIONS

PHASE II
RECOVERY

CREW RETRIEVAL

MQF

PHASE III
LRL

SAMPLE
CREW
SPACECRAFT

RELEASE

LRL

APOLLO BACK CONTAMINATION PROGRAM

Prior to reentering the LM after lunar surface exploration, the crewmen will brush any lunar surface dust or dirt from the space suit using the suit gloves. They will scrape their overboots on the LM footpad and while ascending the LM ladder dislodge any clinging particles by a kicking action.

After entering the LM and pressurizing the cabin, the crew will doff their portable life support system, oxygen purge system, lunar boots, EVA gloves, etc.

The equipment shown in Table I as jettisoned equipment will be assembled and bagged to be subsequently left on the lunar surface. The lunar boots, likely the most contaminated items, will be placed in a bag as early as possible to minimize the spread of lunar particles.

Following LM rendezvous and docking with the CM, the CM tunnel will be pressurized and checks made to insure that an adequate pressurized seal has been made. During this period, the LM, space suits, and lunar surface equipment will be vacuumed. To accomplish this, one additional lunar orbit has been added to the mission.

The lunar module cabin atmosphere will be circulated through the environmental control system suit circuit lithium hydroxide (Li-OH) canister to filter particles from the atmosphere. A minimum of five hours weightless operation and filtering will reduce the original airborne contamination to about 10^{-15} per cent.

To prevent dust particles from being transferred from the LM atmosphere to the CM, a constant flow of 0.8 lbs/hr oxygen will be initiated in the CM at the start of combined LM/CM operation. Oxygen will flow from the CM into the LM then overboard through the LM cabin relief valve or through spacecraft leakage. Since the flow of gas is always from the CM to the LM, diffusion and flow of dust contamination into the CM will be minimized. After this positive gas flow has been established from the CM, the tunnel hatch will be removed.

The CM pilot will transfer the lunar surface equipment stowage bags into the LM one at a time. The equipment listed in Table I as equipment transferred will then be bagged using the "Buddy System" and transferred back into the CM where the equipment will be stowed. The only equipment which will not be bagged at this time are the crewmen's space suits and flight logs.

TABLE I
LUNAR SURFACE EQUIPMENT - CLEANING AND TRANSFER

ITEM	LOCATION AFTER JETTISON	EQUIPMENT LOCATION AT LUNAR LAUNCH	TRANSFER	LM-CM REMARKS
Jettisoned Equipment:				
Overshoes (In Container)	Lunar surface			
Portable Life Support System	Lunar surface			
Camera	Lunar surface			
Lunar tool tether	Lunar surface			
Spacesuit connector cover	Lunar surface			
Equipment Left in LM:				
EVA tether	RH side stowage container	RH side stowage container		Equipment Brushed prior
EVA visors	Helmet bag	Helmet bag		to stowage for
EVA gloves	Helmet bag	Helmet bag		launch
Purge valve	Interim stowage assy	Interim stowage assy		
Oxygen purge system	Engine cover	Engine cover		
Equipment Transferred to CM:				
Spacesuit	On crew	On crew	Stowed in bag	All equipment to be cleaned
Liquid-cooled garment	On crew	On crew	On crew	by vacuum
Helmet	On crew	On crew	Stowed in bag	brush prior to transfer to CM
Watch	On crew	On crew	On crew	
Lunar grab sample	LH stowage	LH stowage	Stowed in bag	
Lunar sample box	SRC rack	SRC rack	Stowed in bag	
Film magazine	SRC rack	SRC rack	Stowed in bag	

POSITIVE GAS FLOW
FROM CM TO LM AFTER POSTLANDING DOCKING

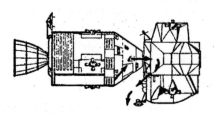

- PROCEDURES

    - PRESSURIZE TUNNEL

    - CM CABIN PRESSURE RELIEF VALVES POSITIONED TO CLOSED

    - LM FORWARD HATCH DUMP/RELIEF VALVE VERIFIED IN AUTOMATIC

    - CM DIRECT O_2 VALVE OPENED TO ESTABLISH CM CABIN PRESSURE
        AT LEAST 0.5 PSI GREATER THAN LM

    - OPEN PRESSURE EQUALIZATION ON TUNNEL HATCH

    - OBSERVE LM CABIN PRESSURE RELIEF FUNCTION

    - ADJUST CM DIRECT O_2 TO STABLE 0.8 #/HR

    - OPEN TUNNEL HATCH

OXYGEN USAGE RATES FOR POSITIVE GAS FLOW
FROM CM TO LM

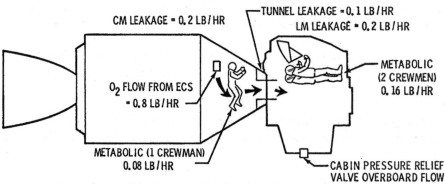

NOMINAL OXYGEN USAGE RATES

CM METABOLIC RATE	0.08 LB/HR
CM LEAKAGE	0.20 LB/HR
TUNNEL LEAKAGE	0.10 LB/HR
LM METABOLIC RATE	0.16 LB/HR
LM LEAKAGE	0.20 LB/HR
FLOW THRU LM CABIN PRESSURE RELIEF VALVE	0.06 LB/HR

Following the transfer of the LM crew and equipment, the spacecraft will be separated and the three crewmen will start the return to Earth. The separated LM contains the remainder of the lunar exposed equipment listed in Table 1.

Command Module Operations — Through the use of operational and housekeeping procedures the command module cabin will be purged of lunar surface and/or other particulate contamination prior to Earth reentry. These procedures start while the LM is docked with the CM and continue through reentry into the Earth's atmosphere.

The LM crewmen will doff their space suits immediately upon separation of the LM and CM. The space suits will be stowed and will not be used again during the trans-Earth phase unless an emergency occurs.

Specific periods for cleaning the spacecraft using the vacuum brush have been established. Visible liquids will be removed by the liquid dump system. Towels will be used by the crew to wipe surfaces clean of liquids and dirt particles. The three ECS suit hoses will be located at random positions around the spacecraft to insure positive ventilation, cabin atmosphere filtration, and avoid partitioning.

During the transearth phase, the command module atmosphere will be continually filtered through the environmental control system lithium hydroxide canister. This will remove essentially all airborne dust particles. After about 63 hours operation essentially none (10^{-90} per cent) of the original contaminates will remain.

Lunar Mission Recovery Operations

Following landing and the attachment of the flotation collar to the command module, the swimmer in a biological isolation garment (BIG) will open the spacecraft hatch, pass three BIGs into the spacecraft, and close the hatch.

The crew will don the BIGs and then egress into a life raft containing a decontaminant solution. The hatch will be closed immediately after egress. Tests have shown that the crew can don their BIGs in less than 5 minutes under ideal sea conditions. The spacecraft hatch will only be open for a matter of a few minutes. The spacecraft and crew will be decontaminated by the swimmer using a liquid agent.

Crew retrieval will be accomplished by helicopter to the carrier and subsequent crew transfer to the Mobile Quarantine Facility. The spacecraft will be retrieved by the aircraft carrier.

Biological Isolation Garment — Biological isolation garment (BIGs), will be donned in the CM just prior to egress and helicopter pickup and will be worn until the crew enters the Mobile Quarantine Facility aboard the primary recovery ship.

The suit is fabricated of a light weight cloth fabric which completely covers the wearer and serves as a biological barrier. Built into the hood area is a face mask with a plastic visor, air inlet flapper valve, and an air outlet biological filter.

Two types of BIGs are used in the recovery operation. One is worn by the recovery swimmer. In this type garment, the inflow air (inspired) is filtered by a biological filter to preclude possible contamination of support personnel. The second type is worn by the astronauts. The inflow gas is not filtered, but the outflow gas (respired) is passed through a biological filter to preclude contamination of the air.

Mobile Quarantine Facility — The Mobile Quarantine Facility, is equipped to house six people for a period up to 10 days. The interior is divided into three sections— lounge area, galley, and sleep/bath area. The facility is powered through several systems to interface with various ships, aircraft, and transportation vehicles. The shell is air and water tight. The principal method of assuring quarantine is to filter effluent air and provide a negative pressure differential for biological containment in the event of leaks.

Non-fecal liquids from the trailer are chemically treated and stored in special containers. Fecal wastes will be contained until after the quarantine period. Items are passed in or out of the MQF through a submersible transfer lock. A complete communications system is provided for intercom and external communications to land bases from ship or aircraft. Emergency alarms are provided for oxygen alerts while in transport by aircraft for fire, loss of power and loss of negative pressure.

Specially packaged and controlled meals will be passed into the facility where they will be prepared in a micro-wave oven. Medical equipment to complete immediate postlanding crew examination and tests are provided.

Lunar Receiving Laboratory — The final phase of the back contamination program is completed in the MSC Lunar Receiving Laboratory. The crew and spacecraft are quarantined for a minimum of 21 days after lunar liftoff and are released based upon the completion of prescribed test requirements and results. The lunar sample will be quarantined for a period of 50 to 80 days depending upon the result of extensive biological tests.

The LRL serves four basic purposes:

The quarantine of the lunar mission crew and spacecraft, the containment of lunar and lunar-exposed materials and quarantine testing to search for adverse effects of lunar material upon terrestrial life.

The preservation and protection of the lunar samples.

The performance of time critical investigations.

The preliminary examination of returned samples to assist in an intelligent distribution of samples to principal investigators.

The LRL has the only vacuum system in the world with space gloves operated by a man leading directly into a vacuum chamber at pressures of 10^{-7} torr. (mm Hg). It has a low level counting facility, whose background count is an order of magnitude better than other known counters. Additionally, it is a facility that can handle a large variety of biological specimens inside Class III biological cabinets designed to contain extremely hazardous pathogenic material.

The LRL covers 83,000 square feet of floor space and includes several distinct areas. These are the Crew Reception Area (CRA), Vacuum Laboratory, Sample Laboratories (Physical and Bio-Science) and an administrative and support area. Special building systems are employed to maintain air flow into sample handling areas and the CRA to sterilize liquid waste and to incinerate contaminated air from the primary containment systems.

The biomedical laboratories provide for the required quarantine tests to determine the effect of lunar samples on terrestrial life. These tests are designed to provide data upon which to base the decision to release lunar material from quarantine.

Among the tests:

a. Germ-free mice will be exposed to lunar material and observed continuously for 21 days for any abnormal changes. Periodically, groups will be sacrificed for pathologic observation.

b. Lunar material will be applied to 12 different culture media and maintained under several environmental conditions. The media will then be observed for bacterial or fungal growth. Detailed inventories of the microbial flora of the spacecraft and crew have been maintained so that any living material found in the sample testing can be compared against this list of potential contaminants taken to the Moon by the crew or spacecraft.

c. Six types of human and animal tissue culture cell lines will be maintained in the laboratory and together with embryonated eggs are exposed to the lunar material. Based on cellular and/or other changes, the presence of viral material can be established so that special tests can be conducted to identify and isolate the type of virus present.

d. Thirty-three species of plants and seedlings will be exposed to lunar material. Seed germination, growth of plant cells or the health of seedlings then observed, and histological, microbiological and biochemical techniques used to determine the cause of any suspected abnormality.

e. A number of lower animals will be exposed to lunar material. These specimens include fish, birds, oysters, shrimp, cockroaches, houseflies, planaria, paramecia and euglena. If abnormalities are noted, further tests will be conducted to determine if the condition is transmissible from one group to another.

The crew reception area provides biological containment for the flight crew and 12 support personnel. The nominal occupancy is about 14 days but the facility is designed and equipped to operate for considerably longer if necessary.

Sterilization And Release Of The Spacecraft

Postflight testing and inspection of the spacecraft is presently limited to investigation of anomalies which happened during the flight. Generally, this entails some specific testing of the spacecraft and removal of certain components of systems for further analysis. The timing of postflight testing is important so that corrective action may be taken for subsequent flights.

The schedule calls for the spacecraft to be returned to port where a team will deactivate pyrotechnics, flush and drain fluid systems (except water). This operation will be confined to the exterior of the spacecraft. The spacecraft will then be flown to the LRL and placed in a special room for storage, sterilization, and postflight checkout.

The Mobile Quarantine Facility being removed from the U.S.S. Hornet.

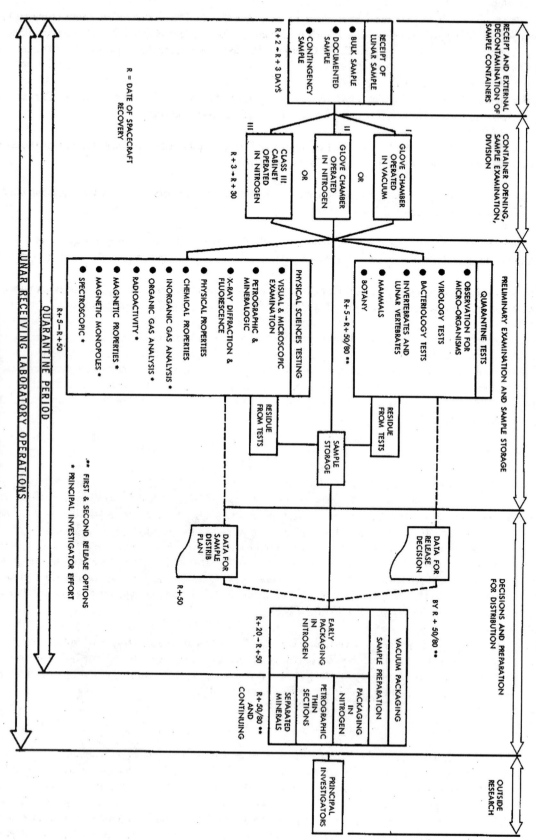

APOLLO PROGRAM MANAGEMENT

The Apollo Program, the United States effort to land men on the Moon and return them safely to Earth before 1970, is the responsibility of the Office of Manned Space Flight (OMSF), National Aeronautics and Space Administration, Washington, D.C. Dr. George E. Mueller is Associate Administrator for Manned Space Flight.

NASA Manned Spacecraft Center (MSC), Houston, is responsible for development of the Apollo spacecraft, flight crew training and flight control. Dr. Robert R. Gilruth is Center Director.

NASA Marshall Space Flight Center (MSFC), Huntsville, Ala., is responsible for development of the Saturn launch vehicles. Dr. Wernher von Braun is Center Director.

NASA John F. Kennedy Space Center (KSC), Fla., is responsible for Apollo/Saturn launch operations. Dr. Kurt H. Debus is Center Director.

The NASA Office of Tracking and Data Acquisition (OTDA) directs the program of tracking and data flow on Apollo 11. Gerald M. Truszynski is Associate Administrator for Tracking and Data Acquisition.

NASA Goddard Space Flight Center (GSFC), Greenbelt, Md., manages the Manned Space Flight Network (MSFN) and Communications Network (NASCOM). Dr. John F. Clark is Center Director.

The Department of Defense is supporting NASA in Apollo 11 during launch, tracking and recovery operations. The Air Force Eastern Test Range is responsible for range activities during launch and down-range tracking. DOD developed jointly with NASA the tracking ships and aircraft. Recovery operations include the use of recovery ships and Navy and Air Force aircraft.

Apollo/Saturn Officials

NASA Headquarters

DR. THOMAS O. PAINE was appointed NASA Administrator March 5, 1969. He was born in Berkeley, Calif., Nov. 9, 1921. Dr. Paine was graduated from Brown University in 1942 with an A.B. degree in engineering. After service as a submarine officer during World War II, he attended Stanford University, receiving an M.S. degree in 1947 and Ph.D. in 1949 in physical metallurgy. Dr. Paine worked as research associate at Stanford from 1947 to 1949 when he joined the General Electric Research Laboratory, Schenectady, N.Y. In 1951 he transferred to the Meter and Instrument Department, Lynn, Mass., as Manager of Materials Development, and later as laboratory manager. From 1958 to 1962 he was research associate and manager of engineering applications at GE's Research and Development Center in Schenectady. In 1963-68 he was manager of TEMPO, GE's Center for Advanced Studies in Santa Barbara, Calif.

On January 31, 1968, President Johnson appointed Dr. Paine Deputy Administrator of NASA, and he was named Acting Administrator upon the retirement of Mr. James E. Webb on Oct. 8, 1968. His nomination as Administrator was announced by President Nixon on March 5, 1969; this was confirmed by the Senate on March 20, 1969. He was sworn in by Vice President Agnew on April 3, 1969.

***

LIEUTENANT GENERAL SAMUEL C. PHILLIPS director of the United States Apollo Lunar Landing Program, was born in Arizona in 1921 and at an early age he moved to Cheyenne, Wyoming which he calls his permanent home. He graduated from the University of Wyoming in 1942 with a B.S. degree in electrical engineering and a presidential appointment as a second lieutenant of infantry in the regular army. He transferred to the Air Corps and earned his pilot's wings in 1943. Following wartime service as a combat pilot in Europe, he studied at the University of Michigan where he received his master of science degree in electrical engineering in 1950. For the next six years he specialized in research and development work at the

Air Material Command, Wright Patterson AFB, Ohio. In June 1956 he returned to England as Chief of Logistics for SAC's 7th Air Division where he participated in writing the international agreement with Great Britain on the use of the Thor IBM. He was assigned to the Air Research and Development Command in 1959 and for four years he was director of the Minuteman program. General Phillips was promoted to Vice Commander of the Ballistic Systems Division in August 1963, and in January 1964 he moved to Washington to become deputy director of the Apollo program. His appointment as Director of the Apollo program came in October of that year.

***

GEORGE H. HAGE was appointed Deputy Director, Apollo Program, in January 1968, and serves as "general manager" assisting the Program Director in the management of Apollo developmental activities. In addition he is the Apollo Mission Director.

Hage was born in Seattle Washington, Oct. 7, 1925, and received his bachelor's degree in electrical engineering from the University of Washington in 1947. He joined Boeing that year and held responsible positions associated with the Bomarc and Minuteman systems, culminating in responsibility for directing engineering functions to activate the Cape Kennedy Minuteman Assembly and test complex in 1962. He then took charge of Boeing's unmanned Lunar Reconnaissance efforts until being named Boeing's engineering manager for NASA's Lunar Orbiter Program in 1963.

Hage joined NASA as Deputy Associate Administrator for Space Science and Applications (Engineering) July 5, 1967, and was assigned to the Apollo Program in October 1967 as Deputy Director (Engineering).

***

CHESTER M. LEE, U.S. NAVY (RET.) was appointed Assistant Apollo Mission Director in August 1966. He was born in New Derry, Pa., in 1919. Lee graduated from the U.S. Naval Academy in 1941 with a BS degree in electrical engineering. In addition to normal sea assignments he served with the Directorate of Research and Engineering, Office of Secretary of Defense and the Navy Polaris missile program. Lee joined NASA in August 1965 and served as Chief of Plans, Missions Operations Directorate, OMSF, prior to his present position.

***

COL. THOMAS H. McMULLEN (USAF) has-been Assistant Mission Director Apollo Program, since March 1968. He was born July 4, 1929, in Dayton, Ohio. Colonel McMullen graduated from the U.S. Military Academy in 1951 with a BS degree. He also received an MS degree from the Air Force Institute of Technology in 1964. His Air Force assignments included: fighter pilot, 1951-1953; acceptance test pilot, 1953-1962; development engineer, Gemini launch vehicle program office, 1964-1966; and Air Force Liaison Officer, 25th Infantry Division, 1967. He served in the Korean and Viet Nam campaigns and was awarded several high military decorations.

***

GEORGE P. CHANDLER, JR., Apollo 11 Mission Engineer, Apollo Operations Directorate, OMSF, Hq., was born in Knoxville, Tenn. Sept. 6, 1935. He attended grammar and high schools in that city was graduated from the University of Tennessee with a B.S. degree in electrical engineering in 1957. He received an army ROTC commission and served on active duty 30 months in the Ordnance Corps as a missile maintenance engineer in Germany. From 1960 until 1965 he was associated with Philco Corp. in Germany and in Houston, Texas. He joined the NASA Office of Manned Space Flight in Washington in 1965 and served in the Gemini and Apollo Applications operations offices before assuming his present position in 1967. Chandler was the mission engineer for Apollo 9 and 10.

***

MAJOR GENERAL JAMES W. HUMPHREYS, JR., USAF Medical Corps joined NASA as Director of Space Medicine, on June 1, 1967. He was born in Fredericksburg, Va., on May 28, 1915. Humphreys graduated from the Virginia Military Institute with a BS degree in chemical engineering in 1935 and from the Medical College of Virginia with an MD in 1939. He served as a medical battalion and group commander in the European Theater in World War II and later as military advisor to the Iranian Army. Humphreys was awarded a master of science in surgery from the Graduate school of the University of Colorado in 1951. Prior to his association with NASA, General Humphreys was Assistant Director, USAID Vietnam for Public Health on a two year tour of duty under special assignment by the Department of State.

Manned Spacecraft Center

ROBERT R. GILRUTH, 55, Director, NASA Manned Spacecraft Center. Born Nashwauk, Minn. Joined NACA Langley Memorial Aeronautical Laboratory in 1936 working in aircraft stability and Control. Organized Pilotless Aircraft Research Division for transonic and supersonic flight research, 1945; appointed Langley Laboratory assistant director, 1952; named to manage manned space flight program, later named Project Mercury, 1958; named director of NASA Manned Spacecraft Center, 1961. BS and MS in aeronautical engineering from University of Minnesota; holds numerous honorary doctorate degrees. Fellow of the Institute of Aerospace Sciences, American Rocket Club and the American Astronautical Society. Holder of numerous professional society, industry and government awards and honorary memberships.

***

GEORGE M. LOW, 43, manager, Apollo Spacecraft Program. Born Vienna, Austria. Married to former Mary R. McNamara. Children: Mark S. 17, Diane E. 15, G. David 13, John M. 11, and Nancy A. 6. Joined NACA Lewis Research Center 1949 specializing in aerodynamic heating and boundary layer research; assigned 1958 to NASA Headquarters as assistant director for manned space flight programs, later becoming Deputy Associate Administrator for Manned Space Flight; named MSC Deputy Director 1964; named manager, Apollo Spacecraft Program in April 1967. BS and MS in aeronautical engineering from Rensselaer Polytechnic Institute, Troy, N.Y.

***

CHRISTOPHER C. KRAFT, JR., 45, MSC Director of Flight Operations. Born Phoebus, Va. Married to Former Elizabeth Anne Turnbull of Hampton, Va. Children: Gordon T. 17, and Kristi-Anne 14. Joined NACA Langley Aeronautical Laboratory in 1945 specializing in aircraft stability and control; became member of NASA Space Task Group in 1958 where he developed basic concepts of ground control and tracking of manned spacecraft. Named MSC Director of Flight Operations in November 1963. BS in aeronautical engineering from Virginia Polytechnic Institute, Blacksburg, Va. Awarded honorary doctorates from Indiana Institute of Technology and Parks College of St. Louis University.

***

KENNETH S. KLEINKNECHT, 50, Apollo Spacecraft Program manager for command and service modules. Born Washington, D.C. Married to former Patricia Jean Todd of Cleveland, Ohio. Children: Linda Mae 19, Patricia Ann 17, and Frederick W. 14. Joined NACA Lewis Research Center 1942 in aircraft flight test; transferred to NACA Flight Research Center 1951 in design and development work in advanced research aircraft; transferred to NASA Space Task Group 1959 as technical assistant to the director; named manager of Project Mercury 1962 and on completion of Mercury, deputy manager Gemini Program in 1963; named Apollo Spacecraft Program manager for command and service modules early 1967 after Gemini Program completed. BS in mechanical engineering Purdue University.

***

CARROLL H. BOLENDER, 49, Apollo Spacecraft Program manager for lunar module. Born Clarksville, Ohio. He and his wife, Virginia, have two children— Carol 22 and Robert 13. A USAF Brigadier general assigned to

NASA, Bolender was named lunar module manager in July 1967 after serving as a mission director in the NASA Office of Manned Space Flight. Prior to joining NASA, he was a member of a studies group in the office of the USAF chief of staff and earlier had worked on USAF aircraft and guided missile systems projects. During World War II, he was a night fighter pilot in the North African and Mediterranean theaters. He holds a BS from Wilmington College, Ohio, and an MS from Ohio State University.

***

DONALD K. SLAYTON, 45, MSC Director of Flight Crew Operations. Born Sparta, Wis. Married to the former Marjorie Lunney of Los Angeles. They have a son, Kent 12. Selected in April 1959 as one of the seven original Mercury astronauts but was taken off flight status when a heart condition was discovered. He subsequently became MSC Director of Flight Crew Operations in November 1963 after resigning his commission as a USAF major. Slayton entered the Air Force in 1943 and flew 56 combat missions in Europe as a B-25 pilot, and later flew seven missions over Japan. Leaving the service in 1946, he earned his BS in aeronautical engineering from University of Minnesota. He was recalled to active duty in 1951 as a fighter pilot, and later attended the USAF Test Pilot School at Edwards AFB, Calif. He was a test pilot at Edwards from 1956 until his selection as a Mercury astronaut. He has logged more than 4000 hours flying time — more than half of which are in jet aircraft.

***

CLIFFORD E. CHARLESWORTH, 37, Apollo 11 prime flight director (green team). Born Redwing, Minn. Married to former Jewell Davis, of Mount Olive, Miss. Children: David Alan 8, Leslie Anne 6. Joined NASA Manned Spacecraft Center April 1962. BS in physics from Mississippi College 1958. Engineer with Naval Mine Defense Lab, Panama City, Fla. 1958-60; engineer with Naval Ordnance Lab, Corona, Calif., 1960-61; engineer with Army Ordnance Missile Command, Cape Canaveral, Fla., 1961-62; flight systems test engineer, MSC Flight Control Division, 1962-65; head, Gemini Flight Dynamics Section, FCD, 1965-66; assistant Flight Dynamics Branch chief, FCD, 1966-68.

***

EUGENE F. KRANZ, 35, Apollo 11 flight director (white team) and MSC Flight Control Division chief. Born Toledo, Ohio. Married to former Marta I. Cadena of Eagle Pass, Texas, Children: Carmen 11, Lucy 9, Joan 7, Mark 6, Brigid 5 and Jean 3. Joined NASA Space Task Group October 1960. Supervisor of missile flight test for McDonnell Aircraft 1958-1960. USAF fighter pilot 1955-1958. McDonnell Aircraft flight test engineer 1954-55. BS in aeronautical engineering from Parks College, St. Louis University, 1954. Assigned as flight director of Gemini 3,4,7/6, 3, 9 and 12; Apollo 5, 8 and 9.

***

GLYNN S. LUNNEY, 32, Apollo 11 flight director (black team). Born Old Forge, Pa. Married to former Marilyn Jean Kurtz of Cleveland, Ohio. Children: Jennifer 8, Glynn 6, Shawn 5 and Bryan 3. Joined NACA Lewis Research Center August 1955 as college co-op employee. Transferred to NASA Space Task Group June 1959. Assigned as flight director of Gemini 9, 10, 11 and 12, Apollo 201, 4, 7, 8 and 10. BS in aeronautical engineering from University of Detroit.

***

MILTON L. WINDLER, 37, Apollo 11 flight director (Maroon team). Born Hampton, Va. Married to former Betty Selby of Sherman, Texas. Children: Peter 12, Marion 9 and Cary 7. Joined NACA Langley Research Center June 1954. USAF fighter pilot 1955-58; rejoined NASA Space Task Group December 1959 and assigned to Recovery Branch of Flight Operations Division in development of Project Mercury recovery equipment and techniques. Later became chief of Landing and Recovery Division Operational Test Branch. Named Apollo flight director team April 1968.

***

CHARLES M. DUKE, 33, astronaut and Apollo 11 spacecraft communicator (CapCom). Born Charlotte, N. C. Married to former Dorothy M. Claiborne of Atlanta, Ga. Children: Charles M. 4. Thomas C. 2. Selected as astronaut in April 1966. Has rank of major in USAF, and is graduate of the USAF Aerospace Research Pilot School. Commissioned in 1957 and after completion of flight training, spent three years as fighter pilot at Ramstein Air Base, Germany. BS in naval sciences from US Naval Academy 1957; MS in aeronautics and astronautics from Massachusetts Institute of Technology 1964. Has more than 24 hours flying time, most of which is jet time.

***

RONALD E. EVANS, 35, astronaut and Apollo 11 spacecraft communicator (CapCom). Born St. Francis, Kans. Married to former Janet M. Pollom of Topeka, Kans. Children: Jaime D. 9 and Jon P. 7. Selected an astronaut in April 1966. Has rank of lieutenant commander in U.S. Navy. Was flying combat missions from USS Ticonderoga off Viet Nam when selected for the astronaut program. Combat flight instructor 1961-1962; made two West Pacific aircraft carrier cruises prior to instructor assignment. Commissioned 1957 through University of Kansas Navy ROTC program. Has more than 3000 hours flying time, most of which is in jets. BS in electrical engineering from University of Kansas 1956; MS in aeronautical engineering from US Naval Postgraduate School 1964.

***

BRUCE McCANDLESS 11, 32, astronaut and Apollo 11 spacecraft communicator (CapCom). Born Boston, Mass. Married to former Bernice Doyle of Rahway, N.J. Children: Bruce III 7 and Tracy 6. Selected as astronaut April 1966. Holds rank of lieutenant commander in US Navy. After flight training and earning naval aviator's wings in 1960, he saw sea duty aboard the carriers USS Forrestal and USS Enterprise, and later was assigned as instrument flight instructor at Oceana, Va. Naval Air Station. He has logged almost 2000 hours flying time, most of which is in jets. BS in naval sciences from US Naval Academy (second in class of 899) 1958; MS in electrical engineering from Stanford University 1965; working on PhD in electrical engineering at Stanford.

***

CHARLES A. BERRY, MD, 45, MSC Director of Medical Research and Operations. Born Rogers, Ark. Married to former Adella Nance of Thermal, Calif. Children: Mike, Charlene and Janice. Joined NASA Manned Spacecraft Center July 1962 as chief of Center Medical Operations Office; appointed MSC Director of Medical Research and Operations May 1966. Previously was chief of flight medicine in the office of the USAF Surgeon General 1959-62; assistant chief, then chief of department of aviation medicine at the School of Aviation Medicine, Randolph AFB, Texas 1956-59 and served as Project Mercury aeromedical monitor; Harvard School of Public Health aviation medicine residency 1955-56; base flight surgeon and command surgeon in stateside, Canal Zone and Caribbean Assignments, 1951-1955. Prior to entering the USAF in 1951, Berry interned at University of California; service at San Francisco City and County Hospital and was for three years in general practice in Indio and Coachella, Calif. BA from University of California at Berkeley 1945; MD University of California Medical School, San Francisco, 1947; Master of public health, Harvard School of Public Health, 1956.

***

DR. WILMOT N. HESS, 42, MSC Director of Science and Applications. Born Oberlin, Ohio. Married to former Winifred Lowdermilk. Children: Walter C. 12, Alison L. 11 and Carl E. 9. Joined NASA Goddard Space Flight Center 1961 as chief of Laboratory for Theoretical Studies; transferred to NASA Manned Spacecraft Center 1967 as Director of Science and Applications. Previously leader, Plowshare Division of University of California Lawrence Radiation Laboratory 1959-61; physics instructor Oberlin College 1948-1949; physics instructor Mohawk College 1947. BS in electrical engineering from Columbia University 1946; MA in physics from Oberlin College 1949; and PhD in physics from University of California 1954.

***

DR. P. R. BELL, 56, chief MSC Lunar and Earth Sciences Division and manager of Lunar Receiving Laboratory. Born Fort Wayne, Indiana. Married to the former Mozelle Rankin. One son, Raymond Thomas 27. Joined NASA Manned Spacecraft Center July 1967. Formerly with Oak Ridge National Laboratories in thermonuclear research, instrumentation and plasma physics, 1946-67; MIT Radiation Laboratories in radar systems development, 1941-46; National Defense Research Committee Project Chicago, 1940-41. Holds 14 patents on electronic measurement devices, thermonuclear reactor components. BS in chemistry and doctor of science from Howard College, Birmingham, Ala.

***

JOHN E. McLEAISH, 39, Apollo 11 mission commentator and chief, MSC Public Information Office. Born Houston, Texas. Married to former Patsy Jo Thomas of Holliday, Texas. Children: Joe D. 19, Carol Ann 14, John E. Jr. 14. Joined NASA Manned Spacecraft Center 1962, named Public Information office chief July 1968. Prior to joining NASA McLeaish was a USAF information officer and rated navigator from 1952 to 1962. BA in journalism from University of Houston. Assigned to mission commentary on Gemini 11 and 12 and Apollo 6 and 8.

***

JOHN E. (JACK) RILEY, 44, Apollo 11 mission commentator and deputy chief MSC Public Information Office. Born Trenton, Mo. Married to former, Patricia C. Pray of Kansas City, Kans. Children: Kevin M. 17, Sean P. 15, Kerry E. 13, Brian T. 9 and Colin D. 6, Joined NASA Manned Spacecraft Center Public Information Office April 1963. Assigned to mission commentary on Gemini 9, 10 and 11 and Apollo 7, 9 and 10. PIO liaison with Apollo Spacecraft Program Office. Prior to joining NASA was public relations representative with General Dynamics/Astronautics 1961-63; executive editor, Independence, Mo. Examiner 1959-61; city editor, Kansas City Kansan 1957-59; reporter, Cincinnati, Ohio Times-Star 1957; reporter-copy editor, Kansas City Kansan 1950-57. Served in US Navy in Pacific-Asiatic Theaters 1942-46. BA in journalism University of Kansas.

***

DOUGLAS K. WARD, 29, born Idaho Falls, Idaho. Married to former Susan Diane Sellery of Boulder, Colorado. Children: Edward 7; Elisabeth, 4; and Cristina, 4. Joined NASA Public Affairs Office June 1966. Responsible for news media activities related to engineering and development and administrative operations at MSC. Assigned to mission commentary on Apollo 7, 8, and 10. BA in political science from the University of Colorado. Before joining NASA worked for two years with the U. S. Information Agency, Voice of America, writing and editing news for broadcast to Latin America and served as assistant space and science editor for the VOA news division.

***

(ROBERT) TERRY WHITE, 41, born Denton, Texas. Married to former Mary Louise Gradel of Waco, Texas. Children: Robert Jr., 4, and Kathleen, 2. Joined NASA Manned Spacecraft Center Public Affairs Office April 1963. Was editor of MSC Roundup (house organ) for four years. Assigned to mission commentary on 12 previous Gemini and Apollo missions. BA in journalism from North Texas State University. Prior to joining NASA, was with Employers Casualty Company, Temco Aircraft Corporation, (now LTV), Johnston Printing Company and Ayres Compton Associates, all of Dallas, Texas.

Marshall Space Flight Center

DR. WERNHER VON BRAUN became the director of MSFC when it was created in 1960. As a field center of NASA, the Marshall Center provides space launch vehicles and payloads, conducts related research, and studies advanced space transportation systems. Dr. von Braun was born in Wirsitz, Germany, on March 23, 1912. He was awarded a bachelor of science degree at the age of 20 from the Berlin Institute of Technology. Two years later, he received his doctorate in physics from the University of Berlin. He was technical director of Germany's rocket program at Peenemunde. Dr. von Braun came to the U.S. in 1945, under a contract to the U.S. Army, along with 120 of his Peenemunde colleagues. He directed high altitude firings of V-2 rockets at White Sands Missile Range, N.M. and later became the project director of the Army's guided missile development unit in Fort Bliss. In 1950 he was transferred to Redstone Arsenal, Ala. The Redstone, the Jupiter and the Pershing missile systems were developed by the von Braun team. Current programs include the Saturn IB and the Saturn V launch vehicles for Project Apollo, the nation's manned lunar landing program and participation in the Apollo Applications program.

***

DR. EBERHARD F.M. REES is deputy director, technical, of NASA-Marshall Space Flight Center. Dr. Rees was born April 28, 1908, in Trossingen, Germany. He received his technical education in Stuttgart and at Dresden Institute of Technology. He graduated from Dresden in 1934 with a master of science degree in mechanical engineering. During World War II. Dr. Rees worked at the German Guided Missile Center in Peenemunde. He came to the United States In 1945 and worked in the Ordnance Research and Development, Sub-Office (rocket), at Fort Bliss. In 1950 the Fort Bliss activities were moved to Redstone Arsenal, Ala. Rees, who became an American citizen in 1954, was appointed deputy director of Research and Development of Marshall Space Flight Center in 1960. He held this position until his appointment in 1963 to deputy director, technical.

***

DR. HERMANN K. WEIDNER is the director of Science and Engineering at the Marshall Space Flight Center. Dr. Weidner has had long and varied experience in the field of rocketry. He became a member of the Peenemunde rocket development group in Germany in 1941. In 1945, he came to the United States as a member of the von Braun research and development team. During the years that followed, this group was stationed at Fort Bliss, as part of the Ordnance Research and Development. After the Fort Bliss group was transferred to Huntsville, Dr. Weidner worked with the Army Ballistic Missile Agency at Redstone Arsenal. He was formerly deputy director of the Propulsion and Vehicle Engineering Laboratory. He was also director of propulsion at MSFC. Dr. Weidner received his U.S. citizenship in April of 1955.

***

MAJ. GEN. EDMUND F. O'CONNOR is director of Program Management at NASA-Marshall Space Flight Center. He is responsible for the technical and administrative management of Saturn launch vehicle programs and that portion of the Saturn/Apollo Applications Program assigned to Marshall. O'Connor was born on March 31, 1922 in Fitchburg, Mass. He graduated from West Point in 1943, he has a bachelor of science in both military engineering and in aeronautical engineering. During World War II, O'Connor served in Italy with the 495th Bombardment Group, and held several other military assignments around the world. In 1962 he went to Norton Air Force Base, as deputy director of the Ballistic Systems Division, Air Force Systems Command. He remained in that position until 1964 when he became director of Industrial Operations (now designated Program Management) at Marshall Space Flight Center.

***

LEE B. JAMES is the manager of the Saturn Program Office in Program Management, Marshall Space Flight Center. A retired Army Colonel, he has been in the rocket field since its infancy. He started in 1947 after graduating in one of the early classes of the Army Air Defense School at Fort Bliss. He is also a graduate of West Point and he holds a master's degree from the University of Southern California at Los Angeles. He joined the rocket development team headed by Dr. Wernher von Braun in 1956. When the team was transferred from the Department of Defense to the newly created NASA in 1960, Jaynes remained as director of the Army's Research and Development Division at Redstone Arsenal. In 1961-62, he was transferred to Korea for a one year tour of duty. After the assignment in Korea, he was transferred by the Army to NASA-MSFC. In 1963 he became manager of the Saturn I and IB program. For a year he served in NASA Headquarters as deputy to the Apollo Program manager. He returned in 1968 to manage the Saturn V program.

MATTHEW W. URLAUB is manager of the S-IC stage in the Saturn Program Office at NASA-Marshall Space Flight Center. Born September 23, 1927 in Brooklyn, he is a graduate of Duke University where he earned his bachelor of science degree in mechanical engineering. Urlaub entered the army in 1950 and finished his tour of duty in 1955. During the period of 1952-1953 he completed a one year course at the Ordnance Guided Missile School at Redstone Arsenal. Upon becoming a civilian he became a member of the Army Ballistic Missile Agency's Industrial Division Staff at Redstone Arsenal. Specifically, he was the ABMA senior resident engineer for the Jupiter Program at Chrysler Corporation in Detroit. He transferred to MSFC in 1961. The field in which he specializes is project engineering/ management.

***

ROY E. GODFREY performs dual roles, one as deputy manager of the Saturn program and he is also the S-II stage manager. Born in Knoxville on November 23, 1922, he earned a bachelor of science degree in mechanical engineering at the University of Tennessee. Godfrey served as second lieutenant in the Air Force during WWII and began his engineering career with TVA. In 1953 he was a member of the research and development team at Redstone Arsenal, when he accepted a position with the Ordnance Missile laboratories. When the Army Ballistic Missile Agency was created in 1956 he was transferred to the new agency. He came to Marshall Center in 1962 to become the deputy director of the Quality and Reliability Assurance Laboratory.

***

JAMES C. McCULLOCH is the S-IVB stage project manager in the Saturn Program Office at the NASA-Marshall Space Flight Center. A native of Alabama, he was born in Huntsville on February 27, 1920. McCulloch holds a bachelor of science degree in mechanical engineering from Auburn University, and a master's degree in business administration from Xavier University. Prior to coming to the Marshall Center in 1961, he had been associated with Consolidated - Vultee Aircraft Corp., National Advisory Committee for Aeronautics; Fairchild Engine and Airplane Corp., and General Electric Co.

***

FREDERICH DUERR is the instrument unit manager in the Saturn Program Office at NASA-Marshall Space Flight Center. Born in Munich, Germany, on January 26, 1909, he is a graduate of Luitpold Oberealschule and the Institute of Technology, both in Munich. He holds B.S. and M.S. degrees in electrical engineering. Duerr specializes in the design of electrical network systems for the rocket launch vehicles. Duerr joined Dr. Wernher von Braun's research and development team in 1941 at Peenemunde, and came with the group to the U.S. in 1945. This group, stationed at White Sands, N.M., was transferred to Huntsville in 1950 to form the Guided Missile Development Division of the Ordnance Missile Laboratories at Redstone Arsenal.

***

DR. FRIDTJOF A. SPEER is manager of the Mission Operations Office in Program Management at the NASA-Marshall Space Flight Center. A member of the rocket research and development team in Huntsville since March 1955, Dr. Speer was assistant professor at the Technical University of Berlin and Physics Editor of the Central Chemical Abstract Magazine in Berlin prior to coming to this country. He earned both his master's degree and Ph.D. in physics from the Technical University. From 1943 until the end of the war, he was a member of the Guided Missile Development group at Peenemunde. Dr. Speer was chief of the Flight Evaluation and Operations Studies Division prior to accepting his present position in August 1965. He became a U.S. citizen in 1960.

***

WILLIAM D. BROWN is manager, Engine Program Office in Program Management at MSFC. A native of Alabama, he was born in Huntsville on December 17, 1926. He is a graduate of Joe Bradley High School in Huntsville and attended Athens College and Alabama Polytechnic Institute to earn his bachelor of science degree in chemical engineering. Following graduation from Auburn University in 1951, he returned to Huntsville to accept a position with the Army research and development team at Redstone Arsenal, where he was involved in catalyst development for the Redstone missile. Shortly after the Army Ballistic Missile Agency was activated at Redstone, Brown became a rocket power plant engineer with ABMA. He transferred en masse to the Marshall Space Flight Center when that organization was established in 1960.

Kennedy Space Center

DR. KURT H. DEBUS, Director, Kennedy Space Center, has been responsible for many state of the art advances made in launch technology and is the conceptual architect of the Kennedy Space Center with its mobile facilities suitable for handling extremely large rockets such as the Saturn V. Born in Frankfurt, Germany, in 1908, he attended Darmstadt University where he earned his initial and advanced degrees in mechanical engineering. In 1939, he obtained his engineering doctorate and was appointed assistant professor at the University. During this period he became engaged in the rocket research program at Peenemunde. Dr. Debus came to the United States in 1945 and played an active role in the U.S. Army's ballistic missile development program. In 1960, he was appointed Director of the Launch Operations Directorate, George C. Marshall Space Flight Center, NASA, at Cape Canaveral. He was appointed to his present post in 1962. He brought into being the government/ industry launch force which has carried out more than 150 successful launches, including those of Explorer I, the Free World's first satellite, the first manned launch and the Apollo 8 flight, first manned orbit of the moon.

***

MILES ROSS, Deputy Director, Center Operations, Kennedy Space Center, is responsible for operations related to engineering matters and the conduct of the Center's technical operations. He has held the position since September 1967. Born in Brunswick, N.J., in 1919, he is a graduate of Massachusetts Institute of Technology where he majored in Mechanical Engineering and Engineering Administration. Prior to his assignment at the Kennedy Space Center, Ross was a project manager of the Air Force Thor and Minuteman Missile systems with TRW, Inc. He was later appointed Director of Flight Operations and Manager of Florida Operations for TRW.

***

ROCCO A. PETRONE, Director of Launch Operations, Kennedy Space Center, is responsible for the management and technical direction of preflight operations and integration, test, checkout and launch of all space vehicles, both manned and unmanned. Born in Amsterdam, N.Y., in 1926, he is a 1946 graduate of the U.S. Military Academy and received a Masters Degree in Mechanical Engineering from Massachusetts Institute of Technology in 1951. His career in rocketry began shortly after graduation from MIT when he was assigned to the Army's Redstone Arsenal, Huntsville, Ala. He participated in the development of the Redstone missile in the early 1950's and was detailed to the Army's General Staff at the Pentagon from 1956 to 1960. He came to KSC as Saturn Project Officer in 1960. He later became Apollo Program Manager and was appointed to his present post in 1966.

***

RAYMOND L. CLARK, Director of Technical Support, Kennedy Space Center, is responsible for the management and technical direction of the operation and maintenance of KSC's test and launch complex facilities, ground support equipment and ground instrumentation required to support the assembly, test, checkout and launch of all space vehicles - both manned and unmanned. Born in Sentinel, Oklahoma, in 1924, Clark attended Oklahoma State University and is a 1945 graduate of the U.S. Military Academy with a degree in military science and engineering. He received a master of science degree in aeronautics and guided missiles from the University of Southern California in 1950 and was a senior project officer for the Redstone and

Jupiter missile projects at Patrick AFB from 1954 to 1957. He joined KSC in 1960. Clark retired from the Army with the rank of lieutenant colonel in 1965.

***

G. MERRITT PRESTON, Director of Design Engineering, Kennedy Space Center, is responsible for design of ground support equipment, structures and facilities for launch operations and support elements at the nation's Spaceport. Born in Athens, Ohio, in 1916, he was graduated from Rensselaer Polytechnic Institute in New York with a degree in aeronautical engineering In 1939. He then joined the National Advisory Committee for Aeronautics (NACA) at Langley Research Center, Virginia, and was transferred in 1942 to the Lewis Flight Propulsion Center at Cleveland, Ohio, where he became chief of flight research engineering in 1945. NACA's responsibilities were later absorbed by NASA and Preston played a major role in Project Mercury and Gemini manned space flights before being advanced to his present post in 1967.

***

FREDERIC H. MILLER, Director of Installation Support, Kennedy Space Center, is responsible for the general operation and maintenance of the nation's Spaceport. Born in Toledo, Ohio, in 1911, he claims Indiana as his home state. He was graduated from Purdue University with a bachelor's degree in electrical engineering in 1932 and a master's degree in business administration from the University of Pennsylvania in 1949. He is a graduate of the Industrial College of the Armed Forces and has taken advanced management studies at the Harvard Business School. He entered the Army Air Corps in 1932, took his flight training at Randolph and Kelly Fields, Texas, and held various ranks and positions in the military service before retiring In 1966 as an Air Force major general. He has held-his present post since 1967.

***

REAR ADMIRAL RODERICK O. MIDDLETON, USN, is Apollo Program Manager, Kennedy Space Center, a post he has held since August, 1967. Born in Pomona, Fla., in 1919, he attended Florida Southern College in Lakeland and was graduated from the U.S. Naval Academy in 1937. He served in the South Pacific during World War II and was awarded a master of science degree from Harvard University in 1946. He joined the Polaris development program as head of the Missile Branch in the Navy's Special Project Office in Washington, D.C., and was awarded the Legion of Merit for his role in the Polaris project in 1961. He held a number of command posts, including that as Commanding Officer of the USS Observation Island, Polaris missile test ship, before being assigned to NASA in October 1965.

***

WALTER J. KAPRYAN, Deputy Director of Launch Operations, Kennedy Space Center, was born in Flint, Michigan, in 1920. He attended Wayne University in Detroit prior to entering the Air Force as a First Lieutenant in 1943. Kapryan joined the Langley Research Center, National Advisory Committee for Aeronautics (NACA) in 1947 and the NASA Space Task Group at Langley in March, 1959. He was appointed project engineer for the Mercury Redstone 1 spacecraft and came to the Cape in 1960 with that spacecraft. In 1963, he established and headed the Manned Spacecraft Center's Gemini Program Office at KSC, participating in all 10 manned Gemini flights as well as Apollo Saturn IB and Saturn V missions before advancement to his present post.

***

DR. HANS F. GRUENE, Director, Launch Vehicle Operations, Kennedy Space Center, is responsible for the preflight testing, preparations and launch of Saturn vehicles and operation and maintenance of associated ground support systems. Born in Braunschweig, Germany, in 1910, he earned his degrees in electrical engineering at the Technical University in his hometown. He received his PhD in 1941 and began his career in guided missile work as a research engineer at the Peenemunde Guided Missile Center in 1943. He came to the United States with the Army's Ordnance Research and Development Facility at Fort Bliss, Texas, in

1945 and held a number of management posts at the Marshall Space Flight Center, Huntsville, Ala., before being permanently assigned to NASA's Florida launch site in June 1965.

***

JOHN J. WILLIAMS, Director, Spacecraft Operations, Kennedy Space Center, is responsible to the Director of Launch Operations for the management and technical integration of KSC Operations related to preparation, checkout and flight readiness of manned spacecraft. Born in New Orleans, La., in 1927, Williams was graduated from Louisiana State University with a bachelor of science degree in electrical engineering in 1949. Williams performed engineering assignments at Wright Patterson Air Force Base, Dayton, Ohio, and the Air Force Missile Test Center, Patrick AFB, Florida, before joining NASA in 1959. Williams played important roles in the manned Mercury and Gemini programs before moving to his current post in 1964.

***

PAUL C. DONNELLY, Launch Operations Manager, Kennedy Space Center, is responsible for the checkout of all manned space vehicles, including both launch vehicle and spacecraft. Born in Altoona, Pa., in 1923, Donnelly attended Grove City College in Pennsylvania, the University of Virginia and the U.S. Navy's electronics and guided missile technical schools. Donnelly performed engineering assignments at naval facilities at Chincoteague, Va., and Patuxent Naval Air Station, Md. Prior to assuming his present post, he was Chief Test Conductor for manned spacecraft at Cape Kennedy for the Manned Spacecraft Center's Florida Operations, his responsibilities extending to planning, scheduling and directing all manned spacecraft launch and prelaunch acceptance tests.

***

ROBERT E. MOSER, Chief, Test Planning Office, Launch Operations Directorate, Kennedy Space Center, is responsible for developing and coordinating KSC launch operations and test plans for the Apollo/Saturn programs. Born in Johnstown, Pa., in 1928, Moser regards Daytona Beach, Fla., as his hometown. A 1950 graduate of Vanderbilt University with a degree in electrical engineering, Moser has been associated with the U.S. space program since 1953 and served as test conductor for the launches of Explorer 1, the first American satellite; Pioneer, the first lunar probe; and the first American manned space flight - Freedom 7 with Astronaut Alan B. Shepard aboard.

***

ISOM A. RIGELL, Deputy Director for Engineering, Launch Vehicle Operations, Kennedy Space Center, is responsible for all Saturn V launch vehicle engineering personnel in the firing room during prelaunch preparations and countdown, providing on-site resolution for engineering problems. Born in Slocomb, Ala., in 1923, Rigell is a 1950 graduate of the Georgia Institute of Technology with a degree in electrical engineering. He has played an active role in the nation's space programs since May, 1951.

***

ANDREW J. PICKETT, Chief, Test and Operations Management Office, Directorate of Launch Vehicle Operations, Kennedy Space Center, is responsible for directing the overall planning of Saturn launch vehicle preparation and prelaunch testing and checkout. Born in Shelby County, Ala., Pickett is a 1950 graduate of the University of Alabama with a degree in mechanical engineering. A veteran of well over 100 launches, Pickett began his rocketry career at Huntsville, Ala., in the early 1950s. He was a member of the Army Ballistic Missile Agency launch group that was transferred to NASA in 1960.

***

GEORGE F. PAGE, Chief of the Spacecraft Operations Division, Directorate of Launch Operations, Kennedy Space Center, is responsible for pre-flight checkout operations, countdown and launch of the Apollo spacecraft. Prior to his present assignment, Page was Chief Spacecraft Test Conductor and responsible for prelaunch operations on Gemini and Apollo spacecraft at KSC. Born in Harrisburg, Pa., in 1924, Page is a 1952 graduate of Pennsylvania State University with a bachelor of science degree in aeronautical engineering.

***

GEORGE T. SASSEEN, Chief, Engineering Division, Spacecraft Operations, Kennedy Space Center, is responsible for test planning and test procedure definition for all spacecraft prelaunch operations. Born at New Rochelle, N.Y., in 1928, he regards Weston, Conn., as his home town. A 1949 graduate of Yale University with a degree in electrical engineering, he joined NASA in July 1961. Prior to his present appointment in 1967, he served as Chief, Ground Systems Division, Spacecraft Operations Directorate, KSC. Sasseen's spacecraft experience extends through the manned Mercury, Gemini and Apollo programs.

***

DONALD D. BUCHANAN, Launch Complex 39 Engineering Manager for the Kennedy Space Center Design Engineering Directorate, is responsible for continuing engineering support at Launch Complex 39. He played a key role in the design, fabrication and assembly of such complex mobile structures as the mobile launchers, mobile service structure and transporters. Born in Macon, Ga., in 1922, Buchanan regards Lynchburg, Va., as his home town. He is a 1949 graduate of the University of Virginia with a degree in mechanical engineering. During the Spaceport construction phase, he was Chief, Crawler-Launch Tower Systems Branch, at KSC.

Office of Tracking and Data Acquisition

GERALD M. TRUSZYNSKI, Associate Administrator for Tracking and Data Acquisition, has held his present position since January, 1968, when he was promoted from Deputy in the same office. Truszynski has been involved in tracking, communication, and data handling since 1947, at Edwards, Cal., where he helped develop tracking and instrumentation for the X-1, X-15, and other high speed research aircraft for the National Advisory Committee for Aeronautics (NACA), NASA's predecessor. He directed technical design and development of the 500 mile aerodynamic testing range at Edwards. Truszynski joined NACA Langley Laboratory in 1944, after graduation from Rutgers University with a degree in electrical engineering. He was transferred to Headquarters in 1960, and became Deputy Associate Administrator in 1961. He is a native of Jersey City, N.J.

***

H R BROCKETT was appointed Deputy Associate Administrator for Tracking and Data Acquisition March 10, 1968, after serving five years as director of operations. He began his career with NACA, predecessor of NASA, in 1947, at the Langley Research Center, Hampton, Va., in the instrumentation laboratory. In 1958-59 he was a member of a group which formulated the tracking and ground instrumentation plans for the United States' first round-the-world tracking network for Project Mercury. In 1959, he was transferred to NASA Headquarters as a technical assistant in tracking operations. Brockett was born Nov. 12, 1924, in Atlanta, Neb. He is a graduate of Lafayette College, 1947.

***

NORMAN POZINSKY, Director of Network Support Implementation Division, Office of Tracking and Data Acquisition, has been associated with tracking development since 1959, when he was detailed to NASA as a Marine Corps officer. He assisted in negotiations for facilities in Nigeria, Canada, and other countries for NASA's worldwide tracking network. He retired from the marines in 1963, and remained in his present position. Before joining NASA he was involved in rocket and guided missile development at White Sands, N.M., and China Lake, Cal., and served in 1956-59 as assistant chief of staff, USMC for guided missile systems. Born in New Orleans in 1917, he is a 1937 graduate of Tulane University, the Senior Command and Staff College, and the U.S. Navy Nuclear Weapons School.

***

FREDERICK B. BRYANT, Director of DOD Coordination Division, Office of Tracking and Data Acquisition, has been involved with technical problems of tracking since he joined NASA's Office of Tracking and Data Acquisition as a staff scientist in 1960. He headed range requirements planning until he assumed his present position in September 1964. He has charge of filling technical requirements for tracking ships and aircraft of the Department of Defense in support of NASA flights. Before joining NASA, Bryant spent 21 years as an electronic scientist at the U.S. Navy's David Taylor Model Basin, Carderock, Md. He worked on instrumentation, guidance control, and test programs for ships, submarines, and underwater devices. Bryant received a B.S. degree in electronic engineering in 1937 at Virginia Polytechnic Institute.

***

CHARLES A. TAYLOR became Director of Operations, Communications, and ADP Division, OTDA, when he joined NASA in November 1968. He is responsible for management and direction of operations of OTDA facilities and has functional responsibility for all NASA Automatic Data Processing. He was employed from 1942 to 1955 at NASA Langley Research Center, Hampton, Va., in research instrumentation for high speed aircraft and rockets. He worked for the Burroughs Corp., Paoli, Pa., in 1955-62, and after that for General Electric Co., Valley Forge, Pa., where he had charge of reliability and quality assurance, and managed the NASA Voyager space probe program. Born in Georgia, April 28, 1919, he received a B.S. degree at Georgia Institute of Technology in 1942.

***

PAUL A. PRICE, Chief of Communications and Frequency Management, OTDA, has held his position since he joined the NASA Headquarters staff in 1960. He is responsible for long-range planning and programming stations, frequencies, equipment, and communications links in NASCOM, the worldwide communications network by which NASA supports its projects on the ground and in space flight. Before he came to NASA, Price was engaged in communications and electronics work for 19 years for the Army, Navy, and Department of Defense. Born January 27, 1913, in Pittsburgh, he received his education in the public schools and was graduated from the Pennsylvania State University in 1935, and did graduate work at the University of Pittsburgh.

***

JAMES C. BAVELY, Director of the Network Operations Branch, OTDA, is responsible for operations management of NASA's networks in tracking, communication, command, and data handling for earth satellites, manned spacecraft, and unmanned lunar and deep space probes. Bavely held technical positions in private industry, the Air Force and Navy before joining NASA in 1961, with extensive experience in instrumentation, computer systems, telemetry, and data handling. Born in Fairmont, W.Va. in 1924, he is a 1949 graduate of Fairmont State College, and has since completed graduate science and engineering courses at George Washington University, University of West Virginia and University of Maryland.

***

E.J. STOCKWELL, Program Manager of MSFN Operations, OTDA, has held his present position since April 1962, when he joined the NASA Headquarters staff. Before that he had charge of ground instrumentation at the Naval Air Test Center, Patuxent River, Md. Stockwell was born May 30, 1926 in Howell, Mich. He received his education at Uniontown, Pa., and Fairmont, W.Va. He attended Waynesburg College and earned a B.S. degree in science from Fairmont State College. He is a director of the International Foundation for the Advancement of Telemetry.

***

LORNE M. ROBINSON, MSFN Equipment Program Manager, OTDA, is responsible for new facilities and equipment supporting NASA's manned space flight projects. He has been with OTDA since July 1963. He joined NASA from the Space Division of North American Rockwell, Downey, Cal., where he had been a senior research engineer on manned flight projects for five years. Previously, he was engaged in research at the University of Michigan Research Institute and the Phillips Chemical Co., Dumas, Tex. He was born December 20, 1930, in Detroit. He holds a degree in chemical engineering from Carnegie Institute of Technology (1952) and electrical engineering from the University of Michigan (1958), and completed graduate courses at UCLA.

Goddard Space Flight Center

OZRO N. COVINGTON is the Assistant Director, Manned Flight Support at the Goddard Center. Before joining NASA in June 1961 he was with the U.S. Army Signal Missile Support Agency as Technical Director for fifteen years. He studied electrical engineering at North Texas Agricultural College in Arlington, Texas, before embarking on an extensive career in radar and communications applications and research and development.

**

HENRY F. THOMPSON is Deputy Assistant Director for Manned Flight Support at the Goddard Center. Thompson graduated from the University of Texas with a B.A. degree in 1949 and a B.S. in 1952. His studies included graduate work at Texas Western College in El Paso and at the New Mexico State College, Las Cruces. Before joining NASA in 1959 he was Technical Director of the U.S. Army Electronics Command at White Sands Missile Range, N.M.

**

LAVERNE R. STELTER is chief of the Communications Division at the Goddard Center. Mr. Stelter received his B.S. in electrical engineering from the University of Wisconsin in 1951. After working with the Army Signal Corps, he joined NASA in 1959 as head of the Goddard Communications Engineering Section. In 1961 he was appointed Ground Systems Manager for TIROS weather satellites and was later assigned the same responsibility for Nimbus. He was appointed to his present position In 1963.

**

H. WILLIAM WOOD is head of the Manned Flight Operations Division at the Goddard Center. Before joining NASA he was a group leader at the Langley Research Center with responsibility for implementing the Project Mercury Network. Mr. Wood earned his BSEE degree at the North Carolina State University.

Department of Defense

MAJOR GENERAL DAVID M. JONES is Commander, Air Force Eastern Test Range and Department of Defense Manager for Manned Space Flight Support Operations.

He was born December 18, 1913, at Marshfield., Oregon, and attended the University of Arizona at Tucson from 1932 to 1936. He enlisted in the Arizona National Guard and served one year in the Cavalry prior to entering pilot training in the summer of 1937.

His military decorations include the Legion of Merit, Distinguished Flying Cross with one Oak Leaf Cluster, Air Medal, Purple Heart, Yum Hwei from the Chinese government, and the NASA Exceptional Service Medal with one device.

**

REAR ADMIRAL FRED E. BAKUTIS is Commander, Task Force 130, the Pacific Manned Spacecraft Recovery Force, in addition to his duty assignment as Commander, Fleet Air Hawaii.

He was born November 4, 1912, in Brockton, Massachusetts, and graduated from the U.S. Naval Academy June 6, 1935.

Admiral Bakutis holds the Navy Cross, the Legion of Merit with Combat "V" the Distinguished Flying Cross with Gold Star and the Bronze Star Medal.

***

REAR ADMIRAL PHILIP S. McMANUS is the Navy Deputy to the Department of Defense Manager for Manned Space Flight Support Operations and Commander, Task Force 140, the Atlantic Manned Spacecraft Recovery Force.

He was born in Holyoke, Massachusetts, on July 18, 1919, and was commissioned an ensign in the U.S. Navy following his graduation from the U.S. Naval Academy, in 1942.

Admiral McManus' decorations include the Legion of Merit with combat "V" Navy and Marine Corps Medal; Navy Commendation Medal with combat "V" and two Bronze Stars. His campaign medals include the European-African-Middle Eastern Campaign Medal with three Bronze Campaign Stars and the Asiatic-Pacific Campaign Medal with one Silver and four Bronze Campaign Stars.

***

BRIGADIER GENERAL ALLISON C. BROOKS is the Commander of Aerospace Rescue and Recovery Service (ARRS). He has the major responsibilities for both planned and contingency air recovery operations during Project Apollo.

He was born in Pittsburgh, Pennsylvania, June 26, 1917. General Brooks attended high school in Pasadena, California, and earned his Bachelor of Science degree from the University of California, Berkeley, California, in 1938. He enlisted a year later as a flying cadet in the Air Force and was graduated from Kelly Field in 1940.

General Brooks was awarded the Legion of Merit with one Oak Leaf Cluster, the Distinguished Flying Cross with two Oak Leaf Clusters, the Soldier's Medal, the Bronze Star Medal, the Air Medal with seven Oak Leaf Clusters and the French Croix de Guerre.

***

COLONEL ROYCE G. OLSON is Director, Department of Defense Manned Space Flight Support Office, located at Patrick AFB, Florida. He was born March 24, 1917, and is a native of Illinois, where he attended the University of Illinois.

He is a graduate of the National War College and holder of the Legion of Merit and Air Medal, among other decorations.

2 1 2

2 3

1 3

1 3

2 2 2

2 3 2

2 2 2

2 2 2

2 2 2

2 2 2

2 2 2

2 2 2

Major Apollo/Saturn V Contractors

Contractor	Item
Bellcomm Washington, D.C.	Apollo System Engineering
The Boeing Co. Washington, D. C.	Technical Integration and Evaluation
General Electric-Apollo Support Dept., Daytona Beach Fla.	Apollo Checkout, and Quality and Reliability
North American Rockwell Corp. Space Div., Downey, Calif.	Command and Service Modules
Grumman Aircraft Engineering Corp., Bethpage, N.Y.	Lunar Module
Massachusetts Institute of Technology, Cambridge, Mass.	Guidance & Navigation (Technical Management)
General Motors Corp., AC Electronics Div., Milwaukee, Wis.	Guidance & Navigation (Manufacturing)
TRW Inc. Systems Group Redondo Beach, Calif.	Trajectory Analysis LM Descent Engine LM Abort Guidance System
Avco Corp., Space Systems Div., Lowell, Mass.	Heat Shield Ablative Material
North American Rockwell Corp. Rocketdyne Div. Canoga Park, Calif.	J-2 Engines, F-1 Engines
The Boeing Co. New Orleans	First Stage (SIC) of Saturn V Launch Vehicles, Saturn V Systems Engineering and Integration Ground Support Equipment
North American Rockwell Corp. Space Div. Seal Beach, Calif.	Development and Production or Saturn V Second Stage (S-II)
McDonnell Douglas Astronautics Co. Huntington Beach, Calif.	Development and Production of Saturn V Third Stage (S-IVB)
International Business Machines Federal Systems Div. Huntsville, Ala.	Instrument Unit

Bendix Corp. Navigation and Control Div. Teterboro, N.J.	Guidance Components for Instrument Unit (Including ST-124M Stabilized Platform
Federal Electric Corp.	Communications and Instrumentation Support, KSC
Bendix Field Engineering Corp.	Launch Operations/complex Support, KSC
Catalytic-Dow	Facilities Engineering and Modifications, KSC
Hamilton Standard Division	Portable Life Support System;
United Aircraft Corp. Windsor Locks, Conn.	LM ECS
ILC Industries Dover, Del.	Space Suits
Radio Corp. of America Van Nuys, Calif.	110A Computer - Saturn Checkout
Sanders Associates Nashua, N.H.	Operational Display Systems Saturn
Brown Engineering Huntsville, Ala.	Discrete Controls
Reynolds, Smith and Hill Jacksonville, Fla.	Engineering Design of Mobile Launchers
Ingalls Iron Works Birmingham, Ala.	Mobile Launchers (ML) (structural work)
Smith/Ernst (joint Venture) Tampa, Fla./Washington, D.C.	Electrical Mechanical Portion of MLs
Power Shovel, Inc. Marion, Ohio	Transporter
Hayes International Birmingham, Ala.	Mobile Launcher Service Arms
Bendix Aerospace Systems Ann Arbor, Mich	Early Apollo Scientific Experi ments Package (EASEP)
Aerojet-Gen. Corp, El Monte. Calif.	Service Propulsion System Engine

APOLLO 11 PRINCIPAL INVESTIGATORS AND INVESTIGATIONS OF LUNAR SURFACE SAMPLES

Investigator	Institution	Investigation
Adams, J.B. Co-Investigator: Jones, R. L.	Caribbean Research Inst. St. Thomas, V.I. NASA Manned Spacecraft Center	Visible and Near-Infrared Reflection Spectroscopy of Returned Lunar Sample at CRI Houston, Texas & Lunar Receiving Lab. (LRL)
Adler, I. Co-Investigators.: Walter, L.S. Goldstein, J. I. Philpotts, J.A. Lowman, P.D. French, B. M.	NASA Goddard Space Flight Center, Greenbelt, Md.	Elemental Analysis by Electron Microprobe
Agrell, S.O. Co-Investigator: Muir, I.O.	University Cambridge, England	Broad Mineralogic Studies
Alvarez, L.W. Co-Investigators: Watt, R.D.	University of California, Berkeley, California	Search for Magnetic Monopoles at LRL
Anders, E. Co-Investigators: Keays, R. R. Ganapathy, R. Jeffery, P.M.	University of Chicago, Chicago	a) Determine 14 Elements By Neutron Activation Analysis b) Measure Cosmic Ray Induced A126 Content
Anderson, O. Co-Investigators: Soga, N. Kumazawa, M.	Lamont Geol. Obs. Columbia Univ. Palisades, N.Y.	Measure Sonic Velocity, Thermal Expansivity, Specific Heat, Dielectric Constant, and Index of Refraction
Arnold, J.R. Co-Investigators: Suess, H.E. Bhandari, N. Shedlovsky, J. Honda, M. Lal, D.	Univ. Calif., San Diego La Jolla, Calif.	Determine Cosmic Ray and Solar Particle Activation Effects
Arrhenius, G.O. Co-Investigators: Reid, A. Fitzgerald, R.	Univ. Calif., San Diego La Jolla, Calif.	Determine Microstructure Characteristics and Composition
Barghoorn, E. Co-Investigator: Philpott, D.	Harvard Univ. Cambridge, Mass. NASA Ames Res. Center, Moffett Field, Calif.	Electron Microscopy of Return ed Lunar Organic Samples
Bastin, J. Co-Investigator: Clegg, P.E.	Queen Mary College London, England	Measure Electric Properties and Thermal Conductivity
Bell, P.M. Co-Investigator: Finger, L.	Carnegie Institution of Washington, Washington D.C.	Determine Crystal Structure of Separated Mineral Phases
Biemann, K.	Mass. Inst. Tech. Cambridge, Mass.	Mass Spectrometric Analyses for Organic Matter in Lunar Crust
Birkebak, R.C. Co-Investigators: Cremers, C.J. Dawson, J.P.	Univ. Kentucky Lexington, Ky.	Measure Thermal Radiative Features and Thermal Conductivity
Bowie, S.H.U Co-Investigators: Horne, J.E.T. Snelling, N.J.	Inst. of Geol. Sciences, London England	Determinative Mineralogy for Opaque Materials by Electron Microprobe, Distribution of Radioactive Material by Auto Radiograph, Analysis for Pb, U and Th Isotopes by Mass Spectrometry

Brown, G.M. Co-Investigators: Emeleus, C.H. Holland, J.G. Phillips, R.	Univ. Durham Durham, England	Petrologic Analysis by Standard Methods; Electron Probe Analysis Reflected Light Microscopy
Burlingame, A.L. Co-Investigator: Biemann, K.	Univ. of Calif., Berkeley, Calif. Mass. Inst. Tech. Cambridge, Mass.	Organic Mass Spectrometer Development for LRL
Calvin, M. Co-Investigator: Burlingame, A.L.	Univ. of Calif., Berkeley, Calif.	Study of Lunar Samples by Mass Spectrometry (Computerized) and Other Analytical Instrumentation
Cameron, E.N.	Univ. Wisconsin Madison, Wis.	Determine Structure, Composition Texture, and Phases of Opaque Material by Many Methods
Carter, N.L,	Yale Univ. New Haven, Conn.	Determine Effects of Shock on Lunar Materials Using Optical X-Ray, and Electron Microscopic Methods
Chao, E.C.T. Co-Investigators: James, O.B. Wilcox, R.E. Minkin, J.A.	U.S. Geol. Survey Washington, D.C.	Shocked Mineral Studies by optical, X-Ray and Microprobe Techniques
Clayton, R.N.	Univ. Chicago	Determine Stable Isotope of Oxygen
Cloud, P. Co-Investigator: Philpott, D.	Univ. Calif., Los Angeles NASA Ames Res. Ctr.	Electron Microscopy of Returned Lunar Organic Samples
Collett, L.S. Co-Investigator: Becker, A.	Geol. Survey, Canada	Determine Electrical Conductivity
Compston, W.C. Co-Investigators: Arriens, P.A. Chappell, B.W. Vernon, M.J.	Australian Nat. University, Canberra	Sr and Sr Isotopes By X-Ray Fluorescence and Mass Spectrometry
Dalrymple, G.B. Co-Investigator: Doell, R.R.	U.S. Geol. Survey, Menlo Park, Calif.	Measure Natural & Induced Thermoluminescence to determine History and Environmental Features of Lunar Materials
Davis, R. Co-Investigator: Stoenner, R.W.	Brookhaven Nat. Lab., L.I., New York	Determine Ar37, Ar39 Content
Doell, R. R. Co-Investigators: Gromme, C.S. Senftle, F.	U.S. Geol. Survey Menlo Park, Calif.	Measurement of Magnetic Properties at LRL and USGS Lab oratories, Survey of Remnant Magnetism of Lunar Samples in Vacuum in the LRL
Douglas, J.A.V. Co-Investigators.: Currie, K.L. Dence, M.R. Traill, R.J.	Geol. Survey of Canada Ottawa, Canada	Petrologic, Mineralogic and Textural Studies
Duke, M.B. Co-Investigator: Smith, R.L.	U.S. Geol. Survey, Washington, D.C.	Determine Size Frequency Distribution Physical Properties and Composition of Lunar Materials of Sub-100 Micron Grain Size
Edgington, J.A. Co-Investigator: Blair, I.M.	Queen Mary College, Univ. London Atomic Energy Res. Establishment	Measure Luminescent and Thermo-luminescent Properties Under Proton (147 MEV) Bombardment

Eglinton, G. Co-Investigator: Lovelock, J.E.	Univ. Bristol Bristol, England	To Establish the Precise Nature of Organic Compounds in Lunar Material
Ehmann, W.D. Co-Investigator: Morgan, J.W.	Univ. Kentucky Lexington, Ky.	Analysis for Major Rack Forming Elements using 14 MEV Neutron Activation
Engel, A.E. Co-Investigator: Engel, A.C.J.	Univ. Calif., San Diego La Jolla, Calif.	Wet Chemical Analysis for Major Elements
Epstein, S. Co-Investigator: Taylor, H.P.	Cal. Inst. Tech. Pasadena, Calif.	Determine Content of Stable Isotopes of O, C, H, and Si by Mass Spectrometry
Evans, H.T. Co-Investigators: Barton, P.B., Jr. Roseboom, E.H.	U.S. Geol. Survey, Washington, D.C.	Crystal Structures of Sulfides and Related Minerals
Fields, P.R. Co-Investigators: Hess, D.C. Stevens, C.	Argonne Nat. Lab. Argonne, Ill.	Measure by Mass Spectrometry the Isotopic Abundances of Heavy Elements
Fireman, E.L.	Smithsonian Inst. Astrophysical Obs. Cambridge, Mass.	(a) Measure the Ar37 and Ar39 Content by Mass Spectrometry (b) Determine Tritium Content by Low Level Counting Techniques
Fleischer, R.L. Co-Investigators: Hanneman, R.E. Kasper, J.S. Price, P.B. Walker, R.M.	General Electric Schnectady, N.Y. Washington Univ. St. Louis, Mo.	(a) Measure Structural Defects in Lunar Materials Through Study of optical, Electrical and Mechanical Properties (b) Determine the Effect of Cosmic Radiation on Lunar Samples by Study of Fossil Tracks Resulting from Charged Particles
Fox, S. Co-Investigators: Harada K. Mueller, G.	Univ. Miami Coral Gables, Fla.	Analysis of Organic Lunar Samples for ALPHA Amino Acids and Polymers Thereof
Fredriksson, K. Co-Investigator: Nelen, J.	Smithsonian Inst. Nat. Museum Washington, D.C.	Elemental Analysis by Electron Microprobe
Friedman, I. Co Investigator: O'Neil, J.R.	U.S. Geol. Survey Denver, Colo.	Isotopic Composition of H, D, and Oxygen
Frondel, C. Co-Investigators: Klein, C. Ito, J.	Harvard Univ. Cambridge, Mass.	Broad Studies of the Texture, Composition, and Relationship of Minerals
Gast, P.W.	Lamont Geol. Obs. Columbia Univ., Palisades, N.Y.	Determine Concentration of the Alkali, Alkaline Earth and Lanthanide Elements by Mass Spectrometry
Gay, P. Co-Investigators: Brown, M.G. McKie, D.	Univ. Cambridge Cambridge, England	X-Ray Crystallographic Studies
Geake, J.E. Co-Investigator: Garlick, G.F.J.	Univ. Manchester Manchester, England	Measure Fluorescence Emission and other Excitation Spectra; Optical Polarization; X-Ray Fluorescence; Electron Spin Resonance; Neutron Activation Analysis

Geiss, J Co-Investigators: Eberhardt, P. Grogler, N. Oeschger, H.	Univ. Berne Berne, Switzerland	Measure Rare Gas Content and Cosmic Ray Produced Tritium by Mass Spectrometry
Gold, T.	Cornell Univ. Ithaca, N.Y.	Particle Size Analysis, Photometric Studies of Radiation Effects from Several Types of Rays; Direct Measure of Radioactive Properties; Dielectric Constant and Loss Tangent
Goles, G.G.	Univ. Oregon Eugene, Ore.	Elemental Abundances by Neutron Activation Analysis
Greenman, N.N. Co-Investigators: Cross, H.G.	McDonnell-Douglas Corp. Santa Monica, Calif.	Determine the Luminescence Spectra and Efficiencies of Lunar Material and Compare with Mineral Composition
Grossman, J.J. Co-Investigators: Ryan, J.A. Mukherjee, N.R.	McDonnell-Douglas Corp. Santa Monica, Calif.	Microphysical, Microchemical and Adhesive Characteristics of the Lunar Materials
Hafner, S. Co-Investigator: Virgo, D.	Univ. Chicago Chicago, Ill.	Using Mossbauer and NMR Techniques Measure the Oxidation State of Iron, Radiation Damage and Al, Na, Fe Energy State in Crystals
Halpern, B. Co-Investigator: Hodgson, G.W.	Stanford Univ. Palo Alto, Calif.	Determine Terrestrial & Extra terrestrial Porphyrins in Association with Amino Acid Compounds
Hapke, B.W. Co-Investigators: Cohen, A.J. Cassidy, W.	Univ. Pittsburgh Pittsburgh, Pa.	Determine Effects of Solar Wind on Lunar Material
Haskin, L.A.	Univ. Wisconsin Madison, Wis.	Determine Rare Earth Element Content by Neutron Activation Analysis
Helsley, C.E. Co-Investigators: Burek, P.J. Oetking, P.	Grad. Res. Center of the Southwest Dallas, Texas	Remanent Magnetism Studies
Helz, A.W. Co-Investigator: Annell, C.S.	U.S. Geol. Survey Washington, D.C.	Special Trace Elements by Emission Spectroscopy
Herr, W. Co-Investigators: Kaufhold, J. Skerra, B. Herpers, U.	Univ. Cologne Cologne, Germany	a) Determine Mn53 Content by High Flux Neutron Bombardment b) Determine Age of Lunar Materials Using Fission Track Method c) Measure Thermoluminescence to Determine Effect of Intrinsic Radioactive and Cosmic Ray Particles and Thermal History
Herzenberg, C.L.	Ill. Inst. of Tech. Chicago	Measure the Energy States of the Iron Bearing Minerals and Possible Effects of Cosmic Radiation
Hess, H.H. Co-Investigator: Otolara, G.	Princeton Univ. Princeton, N.J.	Determine Pyroxene Content by X-Ray and Optical Methods
Heymann, D. Co-Investigators: Adams, J.A.A. Fryer, G.E.	Rice Univ. Houston, Texas	Determine Rare Gases and Radio active Isotopes by Mass Spectrometry

Hintenberger, H. Co-Investigator: Begemann, R.	Max Planck Inst. Fur Chemie, Germany Mainz,	a) Abundance and Isotopic Composition of Hydrogen
Begemann, R. Schultz, L. Vilcsek, E. Wanke, H. Wlotzka, A.		b) Measure Concentration and Isotopic Composition of Rare Gases
Voshage, H. Wanke, H. Schultz, L.		c) isotopic Composition of Nitrogen
Hoering, T. Co-Investigator: Kaplan, I.R.	Carnegie Inst., D.C. Univ. Calif., L.A.	Analytical Lunar Sample Analyses for C13/C12 and D/H of Organic Matter
Hurley, P.M. Co-Investigator: Pinson, W.H., Jr.	Mass. Inst. Tech. Cambridge, Mass.	Analyze for Rb, Sr, and their Isotopes
Jedwab, J.	Univ. Libre De Bruxelles Brussels, Belgium	Determine the Morphological, Optical and Petrographic Properties of Magnetite and its Chemical Composition by Electron Microprobe
Johnson, R.D.	NASA Ames Research Ctr.	Analysis of Lunar Sample for Organic Carbon Behind the Barrier System of the LRL
Kaplan, I.R. Co-Investigators: Berger, R. Schopf, J.W.	Univ. Calif., Los Angeles, Calif.	Ratios of Carbon Hydrogen, Oxygen, and Sulphur Isotope Ratios by Mass Spectrometry
Kanamori, H. Co-Investigators: Mizutani, I I. Takeuchi, H.	Univ. Tokyo Japan	Determine Elastic Constants by Shear/Compressions/Wave Velocity
Keil, K. Co-Investigators: Bunch, T.E. Prinz, M. Snetsinger, K.G.	Univ. New Mexico Albuquerque, N.M.	Elemental Analysis and Mineral Phase Studies by Electron Microprobe
King, E.A. Co-Investigators: Morrison, D.A. Greenwood, W.R.	NASA Manned Spacecraft Center	Non-Destructive Mineralogy & Petrology; Analysis of the Fine Size Fraction of Lunar Materials Including Vitreous 10 Phases
Kohman, T. P. Co-Investigator: Tanner, J.T.	Carnegie Inst. of Tech. Pittsburgh, Pa.	Determine Isotopic Abundance of Pb, Sr, Os, TI, Nd, and Ag by Mass Spectrometry
Kuno, H. Co-Investigator: Kushiro, I.	Univ. Tokyo Japan	Petrographic Analysis for Mineral Identification and Chemical Composition
Larochelle, A. Co-Investigator: Schwarz, E.J.	Geol. Survey, Ottawa, Canada	Thermomagnetic, Magnetic Susceptibility and Remanent Magnetism Studies
Lipsky, S.R. Co-Investigators: Horvath C.G. McMurray, W.J.	Yale Univ. New Haven, Conn.	Identification of Organic Compounds in Lunar Material by Means of Gas Chromatography Mass Spectrometry, NMR, High Speed Liquid Chromatography, and Variations on these techniques

The crew of Apollo 11 (l to r) Neil Armstrong, Michael Collins and Edwin Aldrin.

Armstrong wearing full lunar excursion suit during a simulation before the flight. (left) and suiting up on launch day.(below) The crew at the Cape with Apollo 11 waiting behind them. (bottom)

Collins in a simulator of
the Command Module.
(above)

Apollo 11 leaves Florida
July 16th 1969 (right)

Training to disembark
the Apollo spacecraft
while wearing
quarantine garments.
(below)

"This was our first look at the magnificent
machinery which had been behind us until this
point," *Michael Collins.* (above)

Buzz Aldrin on board the Lunar Module "Eagle" during
the outbound flight of Apollo 11(below left). The crew
took few candid photographs of each other during the
voyage. In fact there are none of Collins.

Aldrin listens in to instructions from Mission Control.
(below)

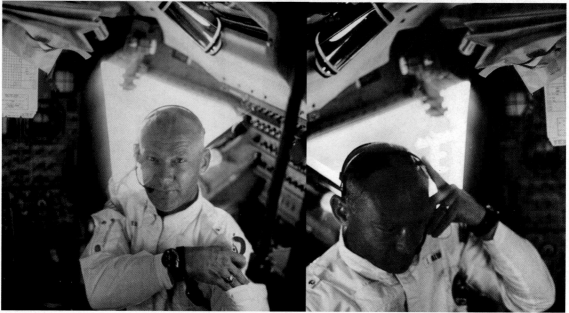

A rarely seen photograph of the Command and Service Module "Columbia". (Above) Most of the color 70mm imagery from Apollo 11 was taken during the EVA on the moon's surface. Meanwhile Command Module pilot Michael Collins ran hundreds of black and white shots of the lunar surface from his vantage point 60 miles above the surface.

The crew sent many excellent television broadcasts during their flight. Armstrong is seen here handling the television camera while standing in the docking tunnel. (left)

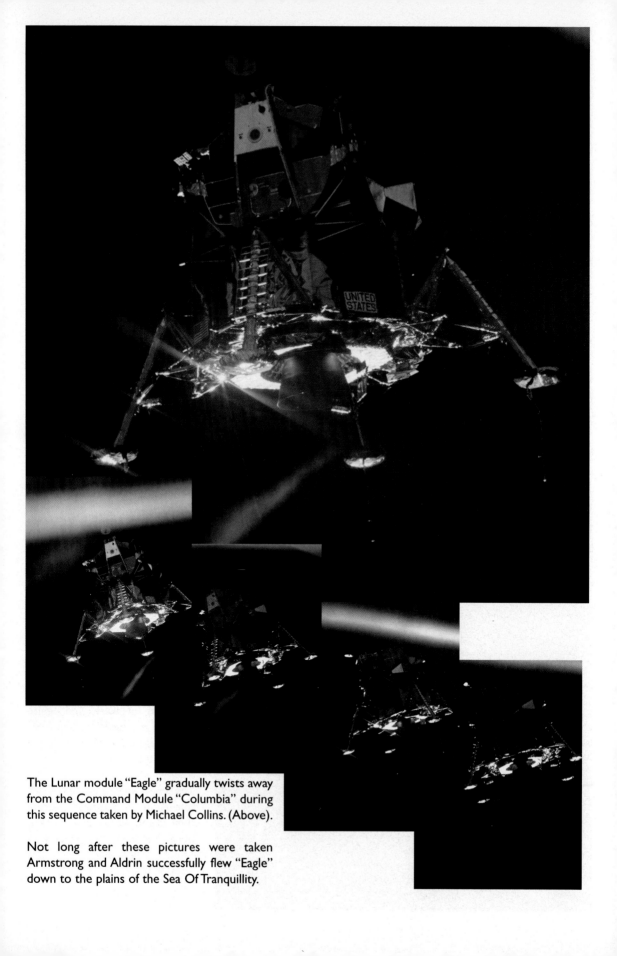

The Lunar module "Eagle" gradually twists away from the Command Module "Columbia" during this sequence taken by Michael Collins. (Above).

Not long after these pictures were taken Armstrong and Aldrin successfully flew "Eagle" down to the plains of the Sea Of Tranquillity.

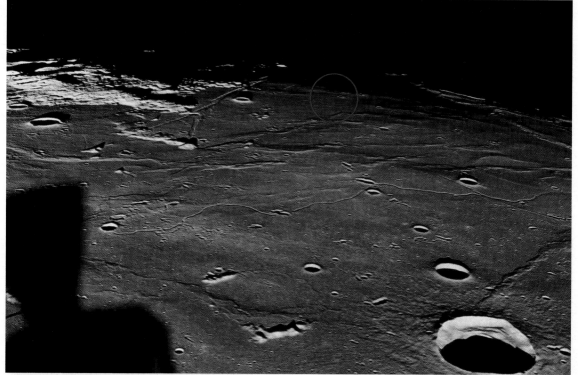

The approach to Tranquillity base. (Top) The shadow in the left foreground is one of the attitude control thrusters on "Eagle". Armstrong manually landed in the area marked on the image.

Aldrin and Armstrong pose with elated smiles after successfully becoming the first humans to land on another celestial body. Ironically because Armstrong was carrying the camera this is one of only three pictures of him while on the moon.

The flight plan called for Armstrong to note how far into the lunar dust the LM had settled. This picture shows one footpad barely making a dent. (left).

The plaque on the front leg of the Lunar Module "Eagle". (above) "Here men from the planet Earth first set foot upon the moon July 1969 AD. We came in peace for all mankind."

This sequence shows astronaut Aldrin descending to the lunar surface (left). The PLSS life support system barely cleared the hatch and so each pilot talked the other through the opening. Aldrin hammers a core sample. (below)

Aldrin and the completely deployed solar wind experiment (above).
Aldrin unloads the rest of the scientific equipment from the back of the LM (below right).
One of the most famous pictures of the 20th century. Armstrong is reflected in Aldrin's gold-plated visor. (below left)

Aldrin and "Old Glory" on the Sea Of Tranquillity. (above)

Aldrin carries the laser reflector and seismometer to an appropriate distance from the LM (above). The deployed seismograph experiment (right) which was used to study the moon's internal structure.

As if on the cue of a movie director the Earth rises above the lunar horizon as "Eagle" returns to dock with "Columbia" (left).

On returning to Earth the crew were transferred to the Mobile Quarantine Facility. (below) Armstrong alleviates the boredom with a banjo. (left)

Home safely the three astronauts share a moment with the media and President Nixon aboard the U.S.S. Hornet from the confines of the Mobile Quarantine Facility. (above)

Investigator	Institution	Analysis
Lovering, J.F. Co-Investigators: Butterfield, D. (a) Kleeman, J.D. (b) Veizer, J. (b) Ware, N.G. (c)	Australian Nat. Univ. Canberra, Australia	a) Neutron Activation for U, Th, K b) Fission Track Analysis for U c) Electron Microprobe Analysis for Elemental Composition
MacGregor, I.D. Co-Investigator: Carter, J.L.	Grad. Research Ctr. S.W. Dallas, Texas	Petrographic Analysis by X-Ray Diffraction and Optical Methods
Manatt, S.L. Co-Investigators: Elleman, D.D. Vaughan, R.W. Chan, S.I.	NASA Jet Propulsion Lab., Pasadena, Calif. Cal. Inst. Tech.	Nuclear Radio Frequency Analysis Including NMR & ESR Analysis for and other Elements and their Chemical State
Mason, B. Co-Investigators: Jarosewich, E. Fredriksson, K. White, J.S.	Smithsonian Inst. Nat. Museum Washington, D.C.	Mineralogic Investigations
Maxwell, J.A. Co-Investigators: Abbey, S. Champ, W.H.	Geol. Survey, Canada Ottawa, Canada	Wet Chemical, X-Ray Fluorescence and Emission Spectroscopy; Flame Photometry for Major/ Minor Elements: Atomic Absorption Spectroscopy
McKay, D.S. Co-Investigators: Anderson, D. H. Greenwood, W.R. Morrison, D.A.	Manned Spacecraft Center	Determine Morphology and Composition of Fine Particles Using Electron Microprobe and Scanning Electron Microscope. Analysis to be Undertaken After Quarantine and not within the LRL
Meinschein, W.G.	Indiana Univ. Bloomington, Ind.	Determine the Alkane C15 To C30 Content By Gas Chromatographic and Mass Spectrometric Techniques
Moore, C.	Arizona State Univ. Temple, Ariz.	Determine Total Concentration of Carbon and Nitrogen
Morrison, G.H.	Cornell Univ.	Elemental Analysis using Spark Source Mass Spectrometry
Muir, A.H., Jr.	North American Rockwell Corp. Science Center, Thousand Oaks, Calif.	Conduct Mossbauer Effect and Spectroscopic Study of Iron Bearing Mineral Separates
Murthy, V.R.	Univ. Minnesota Minneapolis, Minn.	Determine Rare-Earth Elemental and Low Abundance Isotopes of K, Ca, V and Cr Content by Neutron Activation
Nagata, T. Co-Investigators: Ozima, M. Ishikawa, Y.	Univ. Tokyo Japan	Remanent Magnetism Studies
Nagy, B. Co-Investigator: Urey, H.	Univ. Calif., San Diego La Jolla, Calif.	The Presence or Absence of Lipids, Amino Acids, and "Polymer-Type" Organic Matter
Nash, D.B.	NASA Jet Propulsion Lab.	Measure Luminescence, and Physical/Chemical Reaction of Lunar Material to Bombardment by 0.5 to 10 KeV Protons
O'Hara, M.J. Co-Investigator: Biggar, G.M.	Univ. Edinburgh Scotland	Hi-Pressure/Temperature Phase Studies, Determine Temperature of Crystallization of Minerals; Petrologic Studies

O'Kelley, G.D. Co-Investigators: Bell, P.R. Eldridge, J.S. Schonfeld, E. Richardson, K.A.	Oak Ridge Nat. Lab. Tennessee MSC ORNL MSC MSC	Develop the Equipment/Methods for LRL; Measure the K40, U, Th. and Cosmic Ray Induced Radionuclide Content
Oro, J. Co-Investigators: Zlatkis, A. Lovelock, J.E. Becker, R.S. Updegrove,. W. S. Flory, D.A.	Univ. Houston Houston, Texas Manned Spacecraft Ctr.	A Comprehensive Study of the Carbonaceous and Organogenic Matter Present in Returned Lunar Samples with Combination of Gas Chromatographic and Mass Spectrometric Techniques
Oyama, V.I. Co-Investigators: Merek, E. Silverman, M.P.	NASA Ames Res. Ctr. Moffett Field, Calif.	Isolation and Culture of Viable Organisms
Peck, L.C.	U.S. Geol. Survey Denver, Colo.	Standard Wet Chemical Analytical Techniques for Major Elements
Pepin, R.O. Co-Investigator: Nier, A.O.C.	Univ. Minnesota Minneapolis, Minn.	Measure the Elemental and Isotopic Abundances of He, Ne, Ar, Kr, and Xe by Mass Spectrometry
Perkins, R.W. Co-Investigators: Wogman, N.A. Kaye, J.H. Cooper, J.A. Rancitelli, L.A.	Battelle Mem. Inst. Richland, Wash.	Non-Destructive Gamma-Ray Spectrometry for Cosmic Ray Induced and Natural Radio Nuclides
Philpotts, J.A. Co-Investigators: Schnetzler, C. Masuda, A. Thomas, H.H.	NASA Goddard Space Flight Center, Greenbelt, Md.	Determine the Rare Earth Element Content Using Dilution Technique and Mass Spectrometry
Ponnamperuma, C.A. Co-Investigators: Oyama, V.I. Pollack, G. Gehrke, C.W. Zill, L.P.	NASA Ames Res. Center Moffett Field, Calif. Univ. Missouri Ames Res. Center	Analytical Lunar Sample Analyses for Amino Acids, Nucleic Acids, Sugars, Fatty Acids, Hydrocarbons, Porphyrins and Their Components
Quaide, W.L. Co-Investigators: Wrigley, R.C.(a) Debs, R.J. (a) Bunch, T.E. (b)	Ames Res. Center	a) By Non-Destructive Gamma-Ray Spectrometry Determine the Al26, Na22, and Mn54 Content b) Microscopic, X-Ray Diffract ion Analysis to Determine the Effects of Shock on Minerals and Rocks
Ramdohr, P. Co-Investigator; ElGoresy, A.	Max Planck Institut Heidelberg, Germany	Identification of Opaque Minerals, Phases and Composition by X-Ray, Microprobe, and Microscopic Analysis
Reed, G.W. Co-Investigators: Huizenga, J. Jovanovic, S. Fuchs, L.	Argonne Nat. Lab. Argonne, Ill.	Concentration, Isotopic Composition and Distribution of Trace Elements by Neutron Activation Analysis
Reynolds, J.H. Co-Investigators: Rowe, M.W. Hohenberg, C.M.	Univ. Calif., Berkeley	(a) Rare Gas Content by Mass Spectrometry (b) Mass Spectrometry to Ident ify Cosmic Ray Produced Nuclides (c) Mass Spectrometry to Deter mine Rare Gas, K and U Content; Identify Cosmic Ray Produced Nuclides

Rho, J.H. Co-Investigators: Bauman, A.J. Bonner, J.F.	NASA Jet Propulsion Lab. Pasadena, California Cal. Inst. Tech.	Determine Metallic and Non Metallic Porphyrin Content by Fluorescence Spectrophotometry
Richardson, K.A. Co-Investigators: McKay, D.S. Foss, T.H.	NASA Manned Spacecraft Center Houston, Texas	By Autoradiography and Alpha Particle Spectroscopy Identify Alpha Emitting Nuclides
Ringwood, A.E. Co-Investigator: Green, D.H.	Australian Nat'l Univ. Canberra	Petrographic Analysis by Study of Thin and Polished Sections
Robie, R.A. Roedder, E.	U.S. Geol. Survey, D.C. U.S. Geol. Survey, D.C.	Calorimetry (Thermal Properties) Determine the Nature and Composition of Fluid Inclusions, if Present, in Lunar Material
Rose, H.W., Jr. Co-Investigator Cuttitta, F. Dwornik, E.J.	U.S. Geol. Survey, D.C.	X-Ray Fluorescence Methods for Elemental Analysis
Ross, M. Co-Investigators: Warner, J. Papike, J.J. Clark, J.R.	U.S. Geol. Survey Washington, D.C. NASA MSC USGS, D.C.	Determine the Crystallographic Parameters and Composition of Pyroxenes, Micas, Amphiboles, and Host Silicate Minerals by X-Ray Diffraction and Electron Microprobe
Runcorn, S.K.	Univ. of Newcastle Upon Tyne, England	Magnetic Properties in Conjunction with Mineralogic Studies
Schaeffer, O.A. Co-Investigators: Zahringer, J. Bogard, D.	State Univ. of NY. at Stony Brook, N.Y. Max Planck Inst. - Germany Manned Spacecraft Ctr.	Determine Rare Gas Content by Mass Spectrometry at LRL
Schmidt, R.A. Co-Investigator: Loveland, W.D.	Oregon State Univ. Corvallis, Ore.	Determine Rare-Earth and Selected Trace Elements Content by Neutron Activation Analysis. Isotopes of Sm, Eu, Gd will be Determined by Mass Spectrometry
Schopf, W.	Univ. of Calif., L.A. Los Angeles	Micropaleontological Study Using Transmission and Scanning Electron Microscopy
Sclar, C.B. Co-Investigator- Melton, C.W.	Battelle Mem. Inst. Columbus, Ohio	Using Replication and Thin Section Electron Microscopy Determine the Damage in Minerals and Rocks Due to Shock
Scoon, J.H.	Univ. Cambridge England	Wet Chemical Analysis for Major Elements
Short, N.M.	Univ. Houston Houston, Texas	By Petrographic Studies Determine Effect of Shock on Rocks and Minerals and Predict Magnitude of Shock and Line of Impacting Missile
Silver, L.T. Co-Investigator: Patterson, C. C.	Calif. Inst. Tech. Pasadena, Calif.	Determine Lead Isotopes, Concentrations of U, Th, Pb, and their occurrence in Minerals
Simmons, G. Co-Investigators: Brace W.F.) Wones: D.R.) — Parts A & B only	Mass. Inst. Tech. Cambridge, Mass.	a) Calculate Elastic Properties From Measurement of Compression al Shear Wave Velocities at STP b) Measure Thermal Conductivity, Expansion and Diffusivity at STP c) Determine Dielectric Constant, Resistivity Determine Thermal Properties at STP on Samples of Core From Lunar Surface
Sippel, R.F. Co-Investigator: Spencer, A.B.	Mobil Res. and Dev. Corp. Dallas, Texas	Apply Luminescence Petrography to Study of Lunar Materials

Skinner, B.J. Co-Investigator: Winchell, H.	Yale Univ. New Haven, Conn.	Examine the Returned Samples for Condensed Sublimates and if Present Determine the Mineral Phases Present and Elemental Composition
Smales, A.A.	Atomic Energy Research Estab., Harwell, England	Elemental and Isotopic Abundances by Neutron Activation Analysis and by Emission, Spark Source, and X-Ray Fluorescence Spectrography
Smith, J.V. Co-Investigators: Wyllie, P.J. Elders, W.A.	Univ. Chicago Chicago, Ill.	Mineralogic-Petrographic Analysis Using Microprobe, X-Ray Diffraction and Microscopic Methods
Stephens, D.R. Co-Investigator: Keeler, R.N.	Lawrence Radiation Lab. Livermore, California	Physical Properties-Equation of State
Stewart, D.B. Co-Investigators: Appleman, D.E. Papike, J.J. Clark, J.R. Ross, M.	U.S. Geol. Survey, D.C.	Crystal Structure and Stabilities of Feldspars
Strangway, D.W.	Univ. Toronto Canada	Determine Magnetic Properties Including Remanent, Susceptibility, Thermal, Demagnetization, Identify magnetic Minerals
Tatsumoto, M. Co-Investigator: Doe, B.R.	U.S. Geol. Survey Denver, Colo.	Pb Analysis by Mass Spectrometry; U and Th Analysis by Mass Spectrometry and Alpha Spectrometry
Tolansky, S.	Royal Holloway College, Univ. London, England	Isotopic Abundances of U and Th by Microscopic Studies of Diamonds
Turekian, K.K.	Yale Univ. New Haven, Conn.	Determine 20 Elements Having Halflives Greater Than 3 Days By Neutron Activation Analysis
Turkevich, A.L.	Univ. of Chicago Chicago, Ill.	a) Determine Long Lived Isotopes of K, U, and Th by Gamma Ray Spectrometry b) Neutron Activation Analysis for U, Th, Bi, Pb, Tl, and Hg.
Turner, G.	Univ. Sheffield	Determine AR 40 /Ar39 for Age Dating
Urey, H.C. Co-Investigator: Marti, K.	Univ. Calif., S.D. La Jolla, Calif.	Isotopic Abundances by Mass Spectroscopy
Von Engelhardt, W. Co-Investigators: Stoffler, D. Muller, W. Arndt, J.	Univ. Tubingen Tubingen, Germany	Petrographic Study to Determine Shock Effects
Walker, R.M.	Washington Univ. St. Louis, Mo.	a) Measure the Structural Damage to Crystalline Material by Several Techniques b) Geochronological Studies by Investigation of Fission Tracks From Radioactive and Cosmic Ray Particles
Wanke, H. Co-Investigators: Begemann, F. Vilcsek, E. Voshage, H. Begemann, F. Vilcsek, E. Rieder, R. Rieder, R. Wlotzka, F.	Max Planck Inst. Fur Chemie, Mainz Germany	a) Determine K, Th, U Content b) Measure Cosmic Ray Induced Radioactive Nuclides C14 and Cl 36 c) Major Elemental Abundances by Fast Neutron Activation d) Minor Elemental Abundances by Thermal Neutron Activation
Wanless, R.K. Co-Investigators: Stevens, R.D. Loveridge, W.D.	Geol. Survey, Ottawa, Canada	Determine Concentrations of Pb, U, Th, Rb, Sr, Ar, & K and the Isotopic Compositions of Pb and Sr.

Wasserburg, G.J. Co-Investigator: Burnett, D. S.	Cal. Inst. Tech.	Determine K, Ar, Rb, Sr and Rare Gas (He, Ne, Ar, Kr, Xe) Content by Mass Spectrometry
Wasson, J.T. Co-Investigator: Baedecker, P.A.	Univ. Calif., Los Angeles, Calif.	Elemental Abundances for Ga and Ge by Neutron Activation
Weeks, R.A. Co-Investigator: Koloplus, J.	Oak Ridge Nat. Lab. Oak Ridge, Tenn.	Determine the Valence State and Symmetry of the Crystalline Material Using Electron Spin and Nuclear Magnetic Resonance Tech niques and Spin Lattice Relax ation Studies
Weill, D.F.	Univ. Oregon Eugene, Ore.	Determine Temperature of Book Formation by Study of Plagioclase Properties
Wetherill, G.W.	Univ. Calif., Los Angeles, Calif.	Determine Isotopes of Rb. Sr. U, and Pb by Mass Spectrometry
Wiik, H.B. Co-Investigator: Ojanpera, P.M.	Geol. Survey Helsinki, Finland	Wet Chemical Methods to Determine Major Elemental Abundance
Wood, J.A. Co-Investigator: Marvin, U.B.	Smithsonian Inst. Astrophysical Obs. Cambridge, Mass.	Mineralogic and Petrologic Studies by Optical Microscopy, X-Ray Diffraction and Electron Microprobe Measurements
Zahringer, J. Co-Investigators: Kirsten, I. Lammerzahl, P.	Max Planck Inst. Heidelberg Heidelberg, Germany	By Microprobe Analysis and Mass Spectrometry Determine Gas Content and Solar Wind Particle Distribution
Zussman, J.	Univ. Manchester Manchester, England	Geochemical, Mineralogic, and Petrological Studies

A typical sample example of the stereo photographs returned by Armstrong and Aldrin (above left).

The Landing site shown with Houston to scale (left).

APOLLO GLOSSARY

Ablating Materials— Special heat-dissipating materials on the surface of a spacecraft that vaporize during reentry.

Abort— The unscheduled termination of a mission prior to its completion.

Accelerometer— An instrument to sense accelerative forces and convert them into corresponding electrical quantities usually for controlling, measuring, indicating or recording purposes.

Adapter Skirt— A flange or extension of a stage or section that provides a ready means of fitting another stage or section to it.

Antipode— Point on surface of planet exactly 180 degrees opposite from reciprocal point on a line projected through center of body. In Apollo usage, antipode refers to a line from the center of the Moon through the center of the Earth and projected to the Earth surface on the opposite side. The antipode crosses the mid-Pacific recovery line along the 165th meridian of longitude once each 24 hours.

Apocynthion— Point at which object in lunar orbit is farthest from the lunar surface — object having been launched from body other than Moon. (Cynthia, Roman goddess of Moon)

Apogee— The point at which a Moon or artificial satellite in its orbit is farthest from Earth.

Apolune— Point at which object launched from the Moon into lunar orbit is farthest from lunar surface, e.g.: ascent stage of lunar module after staging into lunar orbit following lunar landing.

Attitude— The position of an aerospace vehicle as determined by the inclination of its axes to some frame of reference; for Apollo, an inertial, space-fixed reference is used.

Burnout— The point when combustion ceases in a rocket engine.

Canard— A short, stubby wing-like element affixed to the launch escape tower to provide CM blunt end forward aerodynamic capture during an abort.

Celestial Guidance— The guidance of a vehicle by reference to celestial bodies.

Celestial Mechanics— The science that deals primarily with the effect of force as an agent in determining the orbital paths of celestial bodies.

Cislunar— Adjective referring to space between Earth and the Moon, or between Earth and Moon's orbit.

Closed Loop— Automatic control units linked together with a process to form an endless chain.

Deboost— A retrograde maneuver which lowers either perigee or apogee of an orbiting spacecraft. Not to be confused with deorbit.

Declination— Angular measurement of a body above or below celestial equator, measured north or south along the body's hour circle. Corresponds to Earth surface latitude.

Delta V— Velocity change.

Digital Computer— A computer in which quantities are represented numerically and which can be used to solve complex problems.

Down-Link— The part of a communication system that receives, processes and displays data from a spacecraft.

Entry Corridor— The final flight path of the spacecraft before and during Earth reentry.

Ephemeris— Orbital measurements (apogee, perigee, inclination, period, etc.) of one celestial body in relation to another at given times. In spaceflight, the orbital measurements of a spacecraft relative to the celestial body about which it orbited.

Escape Velocity— The speed a body must attain to overcome a gravitational field, such as that of Earth; the velocity of escape at the Earth's surface is 36,700 feet-per-second.

Explosive Bolts— Bolts destroyed or severed by a surrounding explosive charge which can be activated by an electrical impulse.

Fairing— A piece, part or structure having a smooth, streamlined outline, used to cover a nonstreamlined object or to smooth a junction.

Flight Control System— A system that serves to maintain attitude stability and control during flight.

Fuel Cell— An electrochemical generator in which the chemical energy from the reaction of oxygen and a fuel is converted directly into electricity.

g or g Force— Force exerted upon an object by gravity or by reaction to acceleration or deceleration, as in a change of direction: one g is the measure of force required to accelerate a body at the rate of 32.16 feet-per-second.

Gimbaled Motor— A rocket motor mounted on gimbal; i.e. on a contrivance having two mutually perpendicular axes of rotation, so as to obtain pitching and yawing correction moments.

Guidance System— A system which measures and evaluates flight information, correlates this with target data, converts the result into the conditions necessary to achieve the desired flight path, and communicates this data in the form of commands to the flight control system.

Heliocentric— Sun-centered orbit or other activity which has the Sun at its center.

Inertial Guidance— Guidance by means of the measurement and integration of acceleration from on board the spacecraft. A sophisticated automatic navigation system using gyroscopic devices, accelerometers etc., for high-speed vehicles. It absorbs and interprets such data as speed, position, etc., and automatically adjusts the vehicle to a predetermined flight path. Essentially, it knows where it's going and where it is by knowing where it came from and how it got there. It does not give out any radio frequency signal so it cannot be detected by radar or jammed.

Injection— The process of boosting a spacecraft into a calculated trajectory.

Insertion— The process of boosting a spacecraft into an orbit around the Earth or other celestial bodies.

Multiplexing— The simultaneous transmission of two or more signals within a single channel. The three basic methods of multiplexing involve the separation of signals by time division, frequency division and phase division.

Optical Navigation— Navigation by sight, as opposed to inertial methods, using stars or other visible objects an reference.

Oxidizer— In a rocket propellant, a substance such as liquid oxygen or nitrogen tetroxide which supports combustion of the fuel.

Penumbra— Semi-dark portion of a shadow in which light is partly out off, e.g.: surface of Moon or Earth away from Sun where the disc of the Sun is only partly obscured.

Pericynthion— Point nearest Moon of object in lunar orbit— object having been launched from body other than Moon.

Perigee— Point at which a Moon or an artificial satellite in its orbit is closest to the Earth.

Perilune— The point at which a satellite (e.g.: a spacecraft) in its orbit is closest to the Moon. Differs from pericynthion in that the orbit is Moon-originated.

Pitch— The movement of a space vehicle about an axis (Y) that is perpendicular to its longitudinal axis.

Reentry— The return of a spacecraft that reenters the atmosphere after flight above it.

Retrorocket— A rocket that gives thrust in a direction opposite to the direction of the object's motion.

Right Ascension— Angular measurement of a body eastward along the celestial equator from the vernal equinox (0 degrees RA) to the hour circle of the body. Corresponds roughly to Earth surface longitude, except as expressed in hrs:min:sec instead of 180 degrees west and east from 0 degrees (24 hours=360 degrees).

Roll— The movements of a space vehicle about its longitudinal (X) axis.

S-Band— A radio-frequency band of 1,550 to 5,200 megahertz.

Selenographic— Adjective relating to physical geography of Moon. Specifically, positions on lunar surface as measured in latitude from lunar equator and in longitude from a reference lunar meridian.

Selenocentric— Adjective referring to orbit having Moon as center. (Selene, Gr. Moon)

Sidereal— Adjective relating to measurement of time, position or angle in relation to the celestial sphere and the vernal equinox.

State vector— Ground-generated spacecraft position, velocity and timing information uplinked to the spacecraft computer for crew use as a navigational reference.

Telemetering— A system for taking measurements within an aerospace vehicle in flight and transmitting them by radio to a ground station.

Terminator— Separation line between lighted and dark portions of celestial body which is not self luminous.

Ullage— The volume in a closed tank or container that is not occupied by the stored liquid; the ratio of this volume to the total volume of the tank; also an acceleration to force propellants into the engine pump intake lines before ignition.

Umbra— Darkest part of a shadow in which light is completely absent, e.g.: surface of Moon or Earth away from Sun where the disc of the Sun is completely obscured.

Update pad— Information on spacecraft attitudes, thrust values, event times, navigational data, etc., voiced up to the crew in standard formats according to the purpose, e.g.: maneuver update, navigation check, landmark tracking, entry update, etc.

Up-Link Data— Information fed by radio signal from the ground to a spacecraft.

Yaw— Angular displacement of a space vehicle about its vertical (Z) axis.

APOLLO ACRONYMS AND ABBREVIATIONS

(Note: This list makes no attempt to include all Apollo program acronyms and abbreviations, but several are listed that will be encountered frequently in the Apollo 11 mission. Where pronounced as words in air-to-ground transmissions, acronyms are phonetically shown in parentheses. Otherwise, abbreviations are sounded out by letter.)

AGS	(Aggs)	Abort Guidance System (LM)
AK		Apogee kick
APS	(Apps)	Ascent Propulsion System (LM) -
APS		Auxiliary Propulsion System (S-IVB stage)
BMAG	(Bee-mag)	Body mounted attitude gyro
CDH		Constant delta height
CMC		Command Module Computer
COI		Contingency orbit insertion
CRS		Concentric rendezvous sequence
CSI		Concentric sequence initiate
DAP	(DaPP)	Digital autopilot
DEDA	(Dee-da)	Data Entry and Display Assembly (LM AGS)
DFI		Development flight instrumentation
DOI		Descent orbit insertion
DPS	(Dips)	Descent propulsion system
DSKY	(Diskey)	Display and keyboard
EPO		Earth Parking Orbit
FDAI		Flight director attitude indicator
FITH	(Fith)	Fire in the hole (LM ascent abort staging)
FTP		Full throttle position
HGA		High-gain antenna
IMU		Inertial measurement unit
IRIG	(Ear-ig)	Inertial rate integrating gyro
LOI		Lunar orbit insertion
LPO		Lunar parking orbit
MCC		Mission Control Center
MC&W		Master caution and warning
MSI		Moon sphere of influence
MTVC		Manual thrust vector control
NCC		Combined corrective maneuver
PDI		Powered descent initiation
PIPA	(Pippa)	Pulse integrating pendulous accelerometer
PLSS	(Pliss)	Portable life support system
PTC		Passive thermal control
PUGS	(pugs)	Propellant utilization and gaging system
REFSMMAT	(Refsmat)	Reference to stable member matrix
RHC		Rotation hand controller
RTC		Real-time command
SCS		Stabilization and control system
SHE	(Shee)	Supercritical helium
SLA	(Slah)	Spacecraft LM adapter
SPS		Service propulsion system
TEI		Transearth injection
THC		Thrust hand controller
TIG	(Tigg)	Time at ignition
TLI		Translunar injection
TPF		Terminal phase finalization
TPI		Terminal phase initiate
TVC		Thrust vector control

CONVERSION FACTORS

Multiply	By	To Obtain
Distance:		
feet	0.3048	meters
meters	3.281	feet
kilometers	3281	feet
kilometers	0.6214	statute miles
statute miles	1.609	kilometers
nautical miles	1.852	kilometers
nautical miles	1.1508	statute miles
statute miles	0.86898	nautical miles
statute mile	1760	yards
Velocity:		
feet/sec	0.3048	meters/sec
meters/sec	3.281	feet/sec
meters/sec	2.237	statute mph
feet/sec	0.6818	statute miles/hr
feet/sec	0.5925	nautical miles/hr
statute miles/hr	1.609	km/hr
nautical miles/hr(knots)	1.852	km/hr
km/hr	0.6214	statute miles/hr
Liquid measure, weight:		
gallons	3.785	liters
liters	0.2642	gallons
pounds	0.4536	kilograms
kilograms	2.205	pounds

Multiply	By	To Obtain
Volume:		
cubic feet	0.02832	cubic meters
Pressure:		
pounds/sq inch	70.31	grams/sq cm

Propellant Weights

RP-1 (kerosene) — Approx. 6.7 pounds per gallon

Liquid Oxygen — Approx. 9.5 pounds per gallon

Liquid Hydrogen — Approx. 0.56 pounds per gallon

NOTE: Weight of LH2 will vary as much as plus or minus 5% due to variations in density.

Prelaunch
Mission Operation Report
No. M-932-69-11

TO: A/Administrator   8 July 1969

FROM: MA/Apollo Program Director

SUBJECT: Apollo 11 Mission (AS-506)

No earlier than 16 July 1969, we plan to launch Apollo 11 on the first lunar landing mission. This will be the fourth manned Saturn V flight, the fifth flight of a manned Apollo Command/Service Module, and the third flight of a manned Lunar Module.

Apollo 11 will be launched from Pad A of Launch Complex 39 at the Kennedy Space Center. Lunar touchdown is planned for Apollo Landing Site 2, located in the Southwest corner of the Sea of Tranquillity. The planned lunar surface activities will include collection of a Contingency Sample, assessment of astronaut capabilities and limitations, collection of Bulk Samples, deployment of experiment packages including a laser reflector and instruments for measuring seismic activity, and collection of a Documented Lunar Soil Sample. Photographic records will be obtained and extravehicular activity will be televised. The 8-day mission will be completed with landing in the Pacific Ocean. Recovery and transport of the crew, spacecraft, and lunar samples to the Lunar Receiving Laboratory at the Manned Spacecraft Center will be conducted under quarantine procedures that provide for biological isolation.

Sam C. Phillips
Lt. General, USAF
Apollo Program Director

APPROVAL:

George Mueller
Associate Administrator for
Manned Space Flight

MISSION OPERATION REPORT

APOLLO 11 (AS-506) MISSION

OFFICE OF MANNED SPACE FLIGHT
Prepared by: Apollo Program Office - MAO

Report No. M-932-69-11

HERE MEN FROM THE PLANET EARTH
FIRST SET FOOT UPON THE MOON
JULY 1969, A. D.
WE CAME IN PEACE FOR ALL MANKIND

NEIL A. ARMSTRONG
ASTRONAUT

MICHAEL COLLINS
ASTRONAUT

EDWIN E. ALDRIN, JR.
ASTRONAUT

RICHARD NIXON
PRESIDENT, UNITED STATES OF AMERICA

FOR INTERNAL USE ONLY

MISSION OPERATION REPORT
APOLLO 11 (AS-506) MISSION

OFFICE OF MANNED SPACE FLIGHT
Prepared by: Apollo Program Office - MAO
Report No. M-932-69-11
FOR INTERNAL USE ONLY

FOREWORD

MISSION OPERATION REPORTS are published expressly for the use of NASA Senior Management, as required by the Administrator in NASA Instruction 6-2-10, dated 15 August 1963. The purpose of these reports is to provide NASA Senior Management with timely, complete, and definitive information on flight mission plans, and to establish official mission objectives which provide the basis for assessment of mission accomplishment.

Initial reports are prepared and issued for each flight project just prior to launch. Following launch, updating reports for each mission are issued to keep General Management currently informed of definitive mission results as provided in NASA Instruction 6-2-10.

Because of their sometimes highly technical orientation, distribution of these reports is provided to personnel having program-project management responsibilities. The Office of Public Affairs publishes a comprehensive series of prelaunch and postlaunch reports on NASA flight missions, which are available for general distribution.

APOLLO MISSION OPERATION REPORTS are published in two volumes: the MISSION OPERATION REPORT (MOR); and the MISSION OPERATION REPORT, APOLLO SUPPLEMENT. This format was designed to provide a mission-oriented document in the MOR, with supporting equipment and facility description in the MOR, APOLLO SUPPLEMENT. The MOR, APOLLO SUPPLEMENT is a program oriented reference document with a broad technical description of the space vehicle and associated equipment; the launch complex; and mission control and support facilities.

Published and Distributed by PROGRAM and SPECIAL REPORTS DIVISION (XP)
EXECUTIVE SECRETARIAT - NASA HEADQUARTERS

APOLLO 11 MISSION

The primary purpose of the Apollo 11 Mission is to perform a manned lunar landing and return. During the lunar stay, limited selenological inspection, photography, survey, evaluation, and sampling of the lunar soil will be performed. Data will be obtained to assess the capability and limitations of an astronaut and his equipment in the lunar environment. Figure 1 is a summary of the flight profile.

Apollo 11 will be launched from Pad A of Launch Complex 39 at Kennedy Space Center on 16 July 1969. The Saturn V Launch Vehicle and the Apollo Spacecraft will be the operational configurations. The Command Module (CM) equipment will include a color television camera with zoom lens, a 16mm Maurer camera with 5, 18, and 75mm lenses, and a Hasselblad camera with 80 and 250mm lenses. Lunar Module (LM) equipment will include a lunar television camera with wide angle and lunar day lenses, a 16mm Maurer camera with a 10mm lens, a Hasselblad camera with 80mm lens, a Lunar Surface Hasselblad camera with 60mm lens, and a close-up stereo camera. The nominal duration of the flight mission will be approximately 8 days 3 hours. Translunar flight time will be approximately 73 hours. Lunar touchdown is planned for Landing Site 2, located in the southwest corner of the moon's Sea of Tranquillity. The LM crew will remain on the lunar surface for approximately 21.5 hours. During this period, the crew will accomplish postlanding and pre-ascent procedures and extravehicular activity (EVA).

The nominal EVA plan, as shown in Figure 2, will provide for an exploration period of open-ended duration up to 2 hours 40 minutes with maximum radius of operation limited to 300 feet. The planned lunar surface activities will include in the following order of priority: (1) photography through the LM window, (2) collection of a Contingency Sample, (3) assessment of astronaut capabilities and limitations, (4) LM inspection, (5) Bulk Sample collection, (6) experiment deployment, and (7) lunar field geology including collection of a Documented Lunar Soil Sample. Priorities for activities associated with Documented Sample collection will be: (a) core sample, (b) bag samples with photography, (c) environmental sample, and (d) gas sample. Photographic records will be obtained and EVA will be televised. Assessment of astronaut capabilities and limitations during EVA will include quantitative measurements. There will be two rest and several eat periods.

The total lunar stay time will be approximately 59.5 hours.

The transearth flight time will be approximately 60 hours. Earth landing will be in the Mid-Pacific recovery area with a target landing point located at 172°W longitude and 11° N latitude. Table 1 is a summary of mission events.

Following landing, the flotation collar will be attached to the CM, the CM hatch will be opened and the crew will don Biological Isolation Garments passed in to them by the recovery swimmer. The crew will then egress the CM, transfer to the recovery ship by helicopter, and will immediately enter the Mobile Quarantine Facility (MQF). They will be transported in the MQF to the Lunar Receiving Laboratory (LRL) at the Manned Spacecraft Center. The CM, Sample Return Containers, film, tapes, and astronaut logs will also be transported to the LRL under quarantine procedures.

APOLLO 11 FLIGHT PROFILE

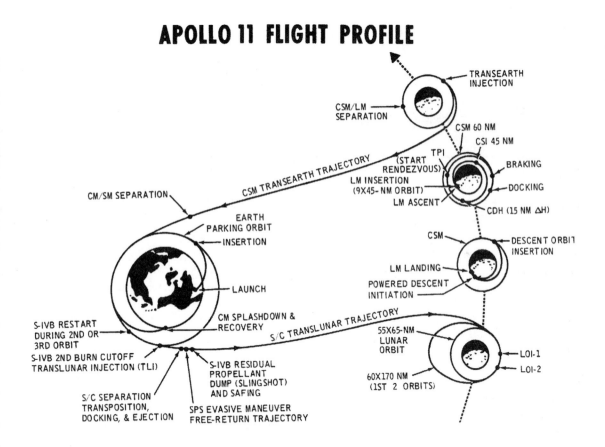

FIGURE 1

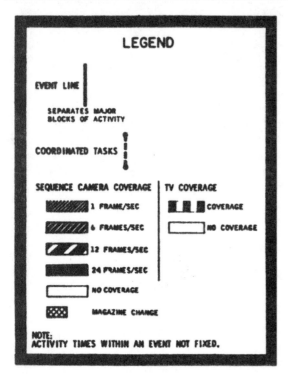

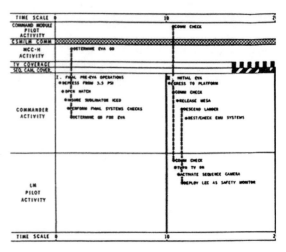

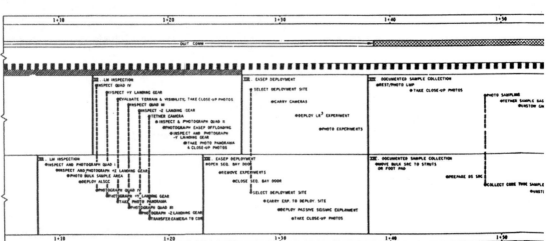

PROGRAM DEVELOPMENT

The first Saturn vehicle was successfully flown on 27 October 1961, initiating operations in the Saturn I Program. A total of 10 Saturn I vehicles (SA-1 to SA-10) was successfully flight tested to provide information on the integration of launch vehicle and spacecraft and to provide operational experience with large multi-engined booster stages (S-1, S-IV).

The next generation of vehicles, developed under the Saturn IB Program, featured an uprated first stage (S-IB) and a more powerful new second stage (S-IVB). The first Saturn IB was launched on 26 February 1966. The first three Saturn IB missions (AS-201, AS-203, and AS-202) successfully tested the performance of the launch vehicle and spacecraft combination, separation of the stages, behavior of liquid hydrogen in a weightless environment, performance of the Command Module heat shield at low earth orbital entry conditions, and recovery operations.

The planned fourth Saturn IB mission (AS-204) scheduled for early 1967 was intended to be the first manned Apollo flight. This mission was not flown because of a spacecraft fire, during a manned prelaunch test, that

M-932-69-11

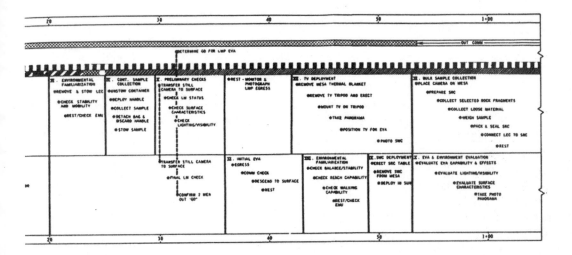

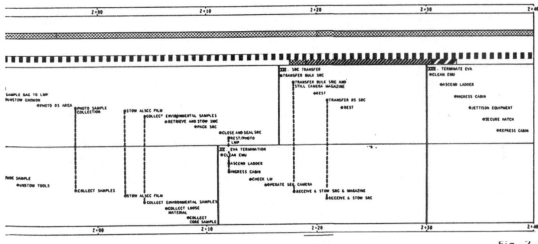

Fig. 2

took the lives of the prime flight crew and severely damaged the spacecraft. The SA-204 Launch Vehicle was later assigned to the Apollo 5 Mission.

The Apollo 4 Mission was successfully executed on 9 November 1967. This mission initiated the use of the Saturn V Launch Vehicle (SA-501) and required an orbital restart of the S-IVB third stage. The spacecraft for this mission consisted of an unmanned Command/Service Module (CSM) and a Lunar Module test article (LTA). The CSM Service Propulsion System (SPS) was exercised, including restart, and the Command Module Block II heat shield was subjected to the combination of high heat load, high heat rate, and aerodynamic loads representative of lunar return entry. All primary mission objectives were successfully accomplished.

The Apollo 5 Mission was successfully launched and completed on 22 January 1968. This was the fourth mission utilizing Saturn IB vehicles (SA-204). This flight provided for unmanned orbital testing of the Lunar Module (LM-1). The LM structure, staging, and proper operation of the Lunar Module Ascent Propulsion System (APS) and Descent Propulsion System (DPS), including restart, were verified. Satisfactory performance of the S-IVB/instrument Unit (IU) in orbit was also demonstrated. All primary objectives were achieved.

TABLE I

MISSION SUMMARY

16 JULY, 72° LAUNCH AZIMUTH FIRST TRANSLUNAR INJECTION OPPORTUNITY

	DURATION (HR:MIN)	GET (DAYS:HR:MIN)	EDT (DAY:HR:MIN
LAUNCH		0:00.00	16:09:32
EARTH ORBIT COAST	2:32		
TRANSLUNAR INJECTION		0:02:44	16:12:16
TRANSLUNAR COAST	73:10		
LUNAR ORBIT INSERTION-1		3:03:54	19:13:26
LUNAR ORBIT INSERTION-2		3:08:09	19:17:41
DESCENT ORBIT INSERTION		4:05:39	20:15:11
LUNAR LANDING		4:06:47	20:16:19
LUNAR STAY	21:36		
EXTRAVEHICULAR ACTIVITY INITIATION		4:16:39	21:20:11
LUNAR EXTRAVEHICULAR ACTIVITY	2:40		
ASCENT		5:04:23	
DOCKING		5:08:00	21:17:32
LM JETTISON		5:11:53	21:21:25
TOTAL LUNAR ORBIT	59:30		
TRANSEARTH INJECTION		5:15:25	22:00:57
TRANSEARTH COAST	59:38		
EARTH LANDING		8:03:17	24:12:49

The Apollo 6 Mission (second unmanned Saturn V) was successfully launched on 4 April 1968. Some flight anomalies were encountered, including oscillations reflecting propulsion-structural longitudinal coupling, an imperfection in the Spacecraft-LM Adapter (SLA) structural integrity, and malfunctions of the J-2 engines in the S-II and S-IVB stages. The spacecraft flew the planned trajectory, but preplanned high velocity reentry conditions were not achieved. A majority of the mission objectives for Apollo 6 was accomplished.

The Apollo 7 Mission (first manned Apollo) was successfully launched on 11 October 1968. This was the fifth and last planned Apollo mission utilizing a Saturn IB Launch Vehicle (SA-205). The 11-day mission provided the first orbital tests of the Block II Command/Service Module. All primary mission objectives were successfully accomplished. In addition, all planned detailed test objectives, plus three that were not originally scheduled, were satisfactorily accomplished.

The Apollo 8 Mission was successfully launched on 21 December and completed on 27 December 1968. This was the first manned flight of the Saturn V Launch Vehicle and the first manned flight to the vicinity of the moon. All primary mission objectives were successfully accomplished. In addition, all detailed test objectives plus four that were not originally scheduled, were successfully accomplished. Ten orbits of the moon were successfully performed, with the last eight at an altitude of approximately 60 NM. Television and photographic coverage was successfully carried out, with telecasts to the public being made in real time.

The Apollo 9 Mission was successfully launched on 3 March and completed on 13 March 1969. This was the second manned Saturn V flight, the third flight of a manned Apollo Command/Service Module, and the first flight of a manned Lunar Module. This flight provided the first manned LM systems performance demonstration. All primary mission objectives were successfully accomplished. All detailed test objectives were accomplished except two associated with S-band and VHF communications which were partially accomplished. The S-IVB second orbital restart, CSM transposition and docking, and LM rendezvous and docking were also successfully demonstrated.

The Apollo 10 Mission was successfully launched on 18 May 1969 and completed on 26 May 1969. This was the third manned Saturn V flight, the second flight of a manned Lunar Module, and the first mission to operate the complete Apollo Spacecraft around the moon. This mission provided operational experience for the crew, space vehicle, and mission-oriented facilities during a simulated lunar landing mission, which followed planned Apollo 11 mission operations and conditions as closely as possible without actually landing. All primary mission objectives and detailed test objectives were successfully accomplished. The manned navigational, visual, and excellent photographic coverage of Lunar Landing Sites 2 and 3 and of the range of possible landing sites in the Apollo belt highlands areas provided detailed support information for Apollo 11 and other future lunar landing missions.

NASA OMSF PRIMARY MISSION OBJECTIVES FOR APOLLO 11

PRIMARY OBJECTIVE

Perform a manned lunar landing and return.

Sam C. Phillips	George E. Mueller
Lt. General, USAF	Associate Administrator for
Apollo Program Director	Manned Space Flight
Date: June 26 1969	Date: June 26, 1969

DETAILED OBJECTIVES AND EXPERIMENTS

The detailed objectives and experiments listed below have been assigned to the Apollo 11 Mission. There are no launch vehicle detailed objectives or spacecraft mandatory and principal detailed objectives assigned to this mission.

	NASA CENTER IDENTIFICATION
Collect a Contingency Sample.	A
Egress from the LM to the lunar surface, perform lunar surface EVA operations, and ingress into the LM from the lunar surface.	B
Perform lunar surface operations with the EMU.	C
Obtain data on effects of DPS and RCS plume impingement on the LM and obtain data on the performance of the LM landing gear and descent engine skirt after touchdown.	D
Obtain data on the lunar surface characteristics from the effects of the LM landing.	E
Collect lunar Bulk Samples.	F
Determine the position of the LM on the lunar surface.	G
Obtain data on the effects of illumination and contrast conditions on crew visual perception.	H
Demonstrate procedures and hardware used to prevent back contamination of the earth's biosphere.	I
Passive Seismic Experiment.	S-031
Laser Ranging Retro-Reflector.	S-078
Solar Wind Composition.	S-080
Lunar Field Geology.	S-059
Obtain television coverage during the lunar stay period.	L
Obtain photographic coverage during the lunar stay period.	M

LAUNCH COUNTDOWN AND TURNAROUND CAPABILITY, AS-506

COUNTDOWN

Countdown (CD) for launch of the AS-506 Space Vehicle (SV) for the Apollo 11 Mission will begin with a precount period starting at T-93 hours during which launch vehicle (LV) and spacecraft (S/C) CD activities will be conducted independently. Official coordinated S/C and LV CD will begin at T-28 hours and will contain two built-in holds; one of 11 hours 32 minutes at T-9 hours, and another of 1 hour at T-3 hours 30 minutes. Figure 3 shows the significant launch CD events.

SCRUB/TURNAROUND

A termination (scrub) of the SV CD could occur at any point in the CD when launch support facilities, SV conditions, or weather warrant. The process of recycling the SV and rescheduling the CD (turnaround) will begin immediately following a scrub. The turnaround time is the minimum time required to recycle and count down the SV to T-0 (liftoff) after a scrub, excluding built-in hold time for launch window synchronization. For a hold that results in a scrub prior to T-22 minutes, turnaround procedures are initiated from the point of hold. Should a hold occur from T-22 minutes (S-II start bottle chilldown) to T-16.2 seconds (S-IC forward umbilical disconnect), then a recycle to T-22 minutes, a hold, or a scrub is possible under the conditions stated in the Launch Mission Rules. A hold between T-16.2 seconds and T-8.9 seconds (ignition) could result in either a recycle or a scrub depending on circumstances. An automatic or manual cutoff after T-8.9 seconds will result in a scrub.

Although an indefinite number of scrub/turnaround cases could be identified, six baseline cases hove been selected to provide the flexibility required to cover probable contingencies. These cases identify the turnaround activities necessary to maintain the same confidence for subsequent launch attempts as for the original attempt. The six cases, shown in Figure 4, are discussed below.

Case 1 - Scrub/Turnaround at Post-LV Cryogenic Loading - Command/Service Module (CSM)/Lunar Module (LM) Cryogenic Reservicing.

Condition: The scrub occurs during CD between T-16.2 and T-8.9 seconds and all SV ordnance items remain connected except the range safety destruct safe and arm (S&A) units. Reservicing of the CSM cryogenics and LM supercritical helium (SHe) is required in addition to the recycling of the LV.

Turnaround Time: Turnaround would require 65 hours consisting of 37 hours for recycle time and 28 hours for countdown time. The time required for a Case 1 turnaround results from flight crew egress, LV cryogenic unloading, LV ordnance operations and battery removal, LM SHe reservicing, CSM cryogenic reservicing, CSM battery removal and installation, and CD resumption at T-28 hours.

Case 2 - Scrub/Turnaround at Post-LV Cryogenic Loading - LM Cryogenic Reservicing

Condition: The scrub occurs during CD between T-16.2 and T-8.9 seconds. Launch vehicle activities are minimized since they fall within allowable time constraints. Reservicing of the LM SHe is required.

Turnaround Time: Turnaround would require 39 hours 15 minutes, consisting of 30 hours 15 minutes for recycle time and 9 hours for CD time. The time requirement for this turnaround case results from flight crew egress, LV cryogenic unloading, LM SHe reservicing, LV loading preparations, and CD resumption at T-9 hours.

Case 3 - Scrub/Turnaround at Post-LV Cryogenic Loading - No CSM/LM Cryogenic Reservicing

Condition: The scrub occurs between T-16.2 and T-8.9 seconds in the CD. Launch vehicle recycle activities are minimized since they fall within allowable time constraints. LM SHe reservicing is not required.

Turnaround Time: Turnaround would require approximately 23 hours 15 minutes, consisting of 14 hours 15 minutes for recycle and 9 hours for CD time. The time required for this case results from flight crew egress, LV cryogenic unloading, S-IC forward umbilical installation and retest, LV propellant preparations, and CD resumption at T-9 hours.

Case 4 - Scrub/Turnaround at Pre-LV Cryogenic Loading - CSM/LM Cryogenic Reservicing

Condition: The scrub occurs at T-8 hours 15 minutes in the CD. The LV requires minimum recycle activities due to the point of scrub occurrence in the CD. The CSM cryogenics require reservicing and the CSM batteries require changing. The LM SHe cryogenics require reservicing. S-II servoactuator inspection is waived.

Turnaround Time: Turnaround would require approximately 59 hours 45 minutes, consisting of 50 hours 45 minutes for recycle and 9 hours for CD. The time required for this turnaround results from CSM cryogenic reservicing, CSM battery removal and installation, LM SHe reservicing, and CD resumption at T-9 hours.

Case 5 - Scrub/Turnaround at Pre-LV Cryogenic Loading - LM Cryogenic Reservicing

Condition: The scrub occurs at T-8 hours 15 minutes in the CD. The SV can remain closed out, except inspection of the S-II servoactuator is waived and the Mobile Service Structure is at the pad gate for reservicing of the LM SHe.

TURNAROUND FROM SCRUB, AS -506

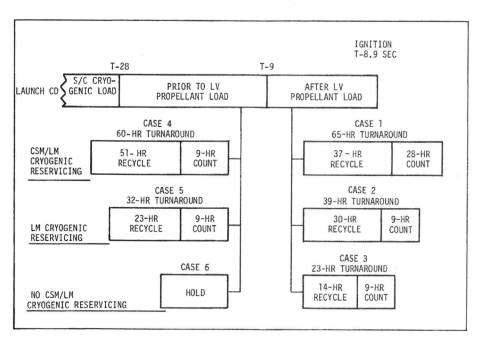

Fig 3

Turnaround Time: Turnaround would require approximately 32 hours consisting of 23 hours for recycle time and 9 hours for CD. This case provides the capability for an approximate 1-day turnaround that exists at T-8 hours 15 minutes in the CD. This capability permits a launch attempt 24 hours after the original T-0. The time required for this turnaround results from LM SHe reservicing and CD resumption at T-9 hours.

Case 6 - Scrub/Turnaround at Pre-LV Cryogenic Loading - No LM/CSM Cryogenic Reservicing

Condition: A launch window opportunity exists 1 day after the original T-0. The LV, LM, and CSM can remain closed out.

Turnaround Time: Hold for the next launch window. The possibility for an approximate 1-day hold may exist at T-8 hours 15 minutes in the CD.

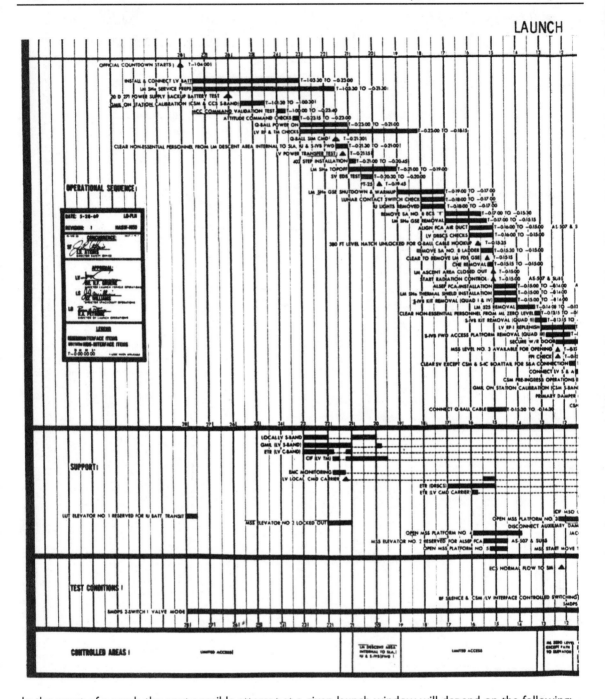

In the event of a scrub, the next possible attempt at a given launch window will depend on the following:

1. The type of scrub/turnaround case occurrence and its time duration.

2. Real-time factors that may alter turnaround time.

3. The number of successive scrubs and the case type of each scrub occurrence.

4. Specific mission launch window opportunities.

Figure 5 shows the scrub/turnaround possibilities in the Apollo 11 Mission for a July launch window. Since the turnaround time may fall short of or exceed a launch window, hold capabilities necessary to reach the

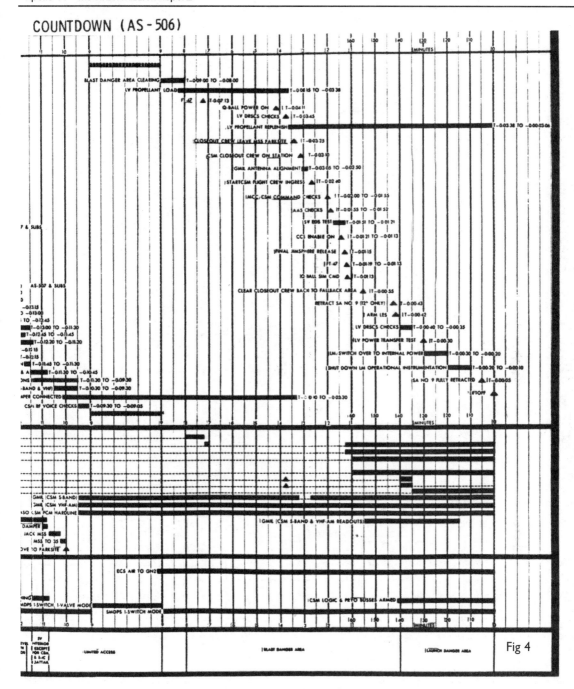

Fig 4

closest possible launch window must be considered. Possible hold points are between recycle and CD, and at T-9 hours in the CD (as in the original CD).

In the event of two successive scrub/turnarounds, SV constraints may require that additional serial or parallel tasks be performed in the second scrub/turnaround case. The 36 possible combinations of the baseline cases and the constraints that may develop on the second turnaround case occurrence are shown in the second scrub/turnaround matrix (Figure 6). A second scrub/turnaround will require that real-time considerations be given either to additional task performance or to task waivers.

SCRUB/TURNAROUND POSSIBILITIES, AS 506
JULY LAUNCH WINDOW

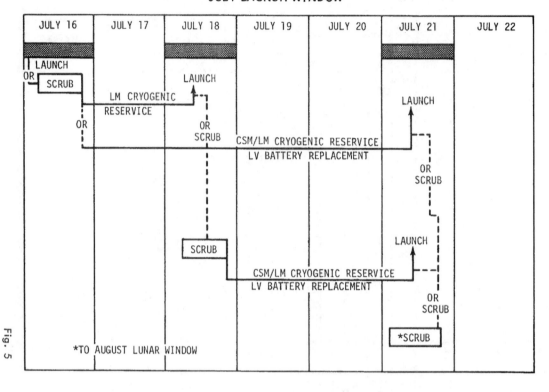

Fig. 5

SECOND SCRUB/TURNAROUND MATRIX, AS-506

		FIRST SCRUB/TURNAROUND					
		CASE 1	CASE 2	CASE 3	CASE 4	CASE 5	CASE 6
SECOND SCRUB/TURNAROUND	CASE 1	YES	YES	YES	YES	YES	YES
	CASE 2	YES	NO A,C,D	NO A,C,D	NO A,C,D	NO A,C,D	NO A,C,D
	CASE 3	YES	NO A,C,D	NO A,B,C	NO D	NO A,C	NO A,B,C
	CASE 4	YES	NO D	NO D	NO D	NO D	NO D
	CASE 5	YES	NO A,C,D	NO A,C	NO D	NO A,C	NO A,C
	CASE 6	YES	NO A,C,D	NO A,B,C	NO D	NO A,C	NO A,B,C

LEGEND

A YES IN THE MATRIX BLOCK INDICATES NO IDENTIFIABLE CONSTRAINTS ARE APPARENT.

A NO FOLLOWED BY ONE OR MORE LETTERS IN THE MATRIX BLOCK INDICATES THAT SOME CONSTRAINT(S), AS IDENTIFIED BELOW, IS APPARENT:

A. THE CSM CRYOGENICS MAY REQUIRE RESERVICING.
B. THE LM SHe MAY REQUIRE RESERVICING.
C. THE CSM BATTERIES MAY REQUIRE CHANGING.
D. THE LV BATTERIES WILL REQUIRE CHANGING.

DETAILED FLIGHT MISSION DESCRIPTION

LAUNCH WINDOWS

Apollo 11 has two types of launch windows. The first, a monthly launch window, defines the days of the month when launch can occur, and the second, a daily launch window, defines the hours of these days when launch can occur.

Monthly Launch Window

Since this mission includes a lunar landing, the flight is designed such that the sun is behind the Lunar Module (LM) and low on the eastern lunar horizon in order to optimize visibility during the LM approach to one of the three Apollo Lunar Landing Sites available during the July monthly launch window. Since a lunar cycle is approximately 28 earth days long, there are only certain days of the month when these landing sites are properly illuminated. Only one launch day is available for each site for each month. Therefore, the Apollo 11 launch must be timed so that the spacecraft will arrive at the moon during one of these days. For a July 1969 launch, the monthly launch window is open on the 16th, 18th, and 21st days of the month. The unequal periods between these dates are a result of the spacing between the selected landing sites on the moon. Table 2 shows the opening and closing of the monthly launch windows and the corresponding sun elevation angles. Figure 7 shows the impact of July launch windows on mission duration.

TABLE 2

MONTHLY LAUNCH WINDOWS

Site	Date	July (EDT) Open-Close**	SEA***	Date	August (EDT) Open-Close**	SEA
2	16	09:32-13:54	10.8°	14H	07:45-12:15	6.0°
3	18H*	11:32-14:02	11.0°	16H	07:55-12:25	6.0°
5	21H	12:09-14:39	9.1°	20H	09:55-14:35	10.0°

*Hybrid (H) trajectory used.

**Based on 108° launch azimuth upper limit.

***Sun Elevation Angle (SEA) - assumes launch at window opening and translunar injection at the first opportunity.

NOTE: A hybrid trajectory is required for a launch on 18 July to make it possible for the Goldstone tracking station 210-foot antenna to cover the LM powered descent phase.

Daily Launch Windows

The maneuver to transfer the S-IVB/spacecraft from earth parking orbit to a translunar trajectory must be performed over a point called the moon's antipode. This is a point on the earth's surface where an imaginary line, drawn from the moon's position (at expected spacecraft arrival time) through the center of the earth, will intersect the far side of the earth. In other words, it is the point on the earth that is exactly opposite the moon. Since the moon revolves around the earth and the earth is spinning on its axis, the antipode is constantly moving. This presents the problem of having the S-IVB/spacecraft rendezvous with a moving target, the

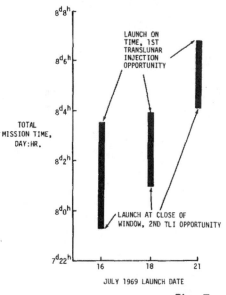

MISSION DURATIONS, JULY LAUNCH WINDOWS

JULY 1969 LAUNCH DATE

Fig. 7

antipode, before it can perform the translunar injection (TLI) burn. Additional constraints on the execution of this maneuver are: (1) it will be performed over the Pacific Ocean, (2) it can occur no earlier than revolution 3 because of S-IVB systems lifetime. These constraints, combined with a single fixed launch azimuth, allow only a very short period of time each day that launch can be performed.

To increase the amount of time available each day, and still maintain the capability to rendezvous with the antipode, a variable launch azimuth technique will be used. The launch azimuth increases approximately 8° per hour during the launch window, and the variation is limited by range safety considerations to between 72° and 106°. This extends the time when rendezvous with the projected antipode can be accomplished up to a maximum of approximately 4.5 hours. The minimum daily launch window for Apollo 11 is approximately 2.5 hours.

FREE-RETURN/HYBRID TRAJECTORY

A circumlunar free-return trajectory, by definition, is one which circumnavigates the moon and returns to earth. The perigee altitude of the return trajectory is of such a magnitude that by using negative lift the entering spacecraft can be prevented from skipping out of the earth's atmosphere, and the aerodynamic deceleration can be kept below 10 g's. Thus, even with a complete propulsion system failure following TLI, the spacecraft would return safely to earth. However, free-return trajectory severely limits the accessible area on the moon because of the very small variation in allowable lunar approach conditions and because the energy of the lunar approach trajectory is relatively high. The high approach energy causes the orbit insertion velocity change requirement (ΔV) to be relatively high.

Since the free-return flight plan is so constraining on the accessible lunar area, hybrid trajectories have been developed that retain most of the safety features of the free return, but do not suffer from the performance penalties. If a hybrid trajectory is used for Apollo 11, the spacecraft will be injected into a highly eccentric elliptical orbit which had the free-return characteristic; i.e., a return to the entry corridor without any further maneuvers. The spacecraft will not depart from the free-return ellipse until spacecraft ejection from the launch vehicle has been completed. After the Service Propulsion System (SPS) has been checked out, a midcourse maneuver will be performed by the SPS to place the spacecraft on a lunar approach trajectory. The resulting lunar approach will not be on a free-return trajectory, and hence will not be subject to the same limitation in trajectory geometry.

On future Apollo lunar missions, landing sites at higher latitudes will be achieved, with little or no plane change, by approaching the moon on a highly inclined trajectory.

LUNAR LANDING SITES

The following Lunar Landing Sites, as shown in Figure 8, are final choices for Apollo 11:

Site 2   latitude 0°41' North — longitude 23°43' East
                Site 2 is located on the east central part of the moon in southwestern
                Mare Tranquillitatis.
Site 3   latitude 0°21' North — longitude 1°18' West
                Site 3 is located near the center of the visible face of the moon in the
                southwestern part of Sinus Medii.
Site 5   latitude 1°41' North — longitude 41°54' West
                Site 5 is located on the west central part of the visible face in southeastern
                Oceanus Procellarum.

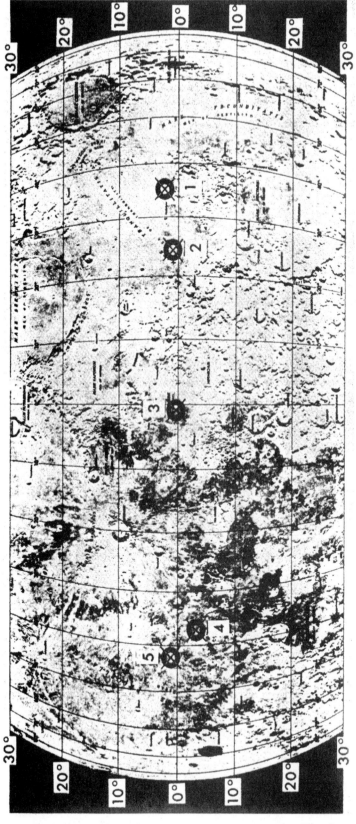

APOLLO LUNAR LANDING SITES

Fig. 8

The final site choices were based on these factors:

Smoothness (relatively few craters and boulders).

Approach (no large hills, high cliffs, or deep craters that could cause incorrect altitude signals to the Lunar Module landing radar).

Propellant requirements (selected sites require the least expenditure of spacecraft propellants).

* Recycle (selected sites allow effective launch preparation recycling if the Apollo/Saturn V countdown is delayed).

* Free-return (sites are within reach of the spacecraft launched on a free-return translunar trajectory).

* Slope (there is little slope - less than 2 degrees in the approach path and landing area).

FLIGHT PROFILE

Launch to Earth Parking Orbit

The Apollo 11 Space Vehicle is planned to be launched at 09:32 EDT from Complex 39A at the Kennedy Space Center, Florida, on a launch azimuth of 72°. The space vehicle (SV) launch weight breakdown is shown in Table 3. The Saturn V boost to earth parking orbit (EPO), shown in

Figure 9, will consist of a full burn of the S-IC and S-II stages and a partial burn of the S-IVB stage of the Saturn V Launch Vehicle. Insertion into a 103-nautical mile (NM) EPO (inclined approximately 33 degrees from the earth's equator) will occur approximately 11.5 minutes ground elapsed time (GET) after liftoff. The vehicle combination placed in earth orbit consists of the SIVB stage, the Instrument Unit (IU), the Lunar Module (LM), the Spacecraft-LM Adapter (SLA), and the Command/Service Module (CSM). While in EPO, the S-IVB and spacecraft will be readied for the second burn of the S-IVB to achieve the translunar injection (TLI) burn. The earth orbital configuration of the SV is shown in Figure 10.

Translunar Injection

The S-IVB J-2 engine will be reignited during the second parking orbit (first opportunity) to inject the SV combination into a translunar trajectory. The second opportunity for TLI will occur on the third parking orbit. The TLI burn will be biased for a small overburn to compensate for the Service Propulsion System (SPS) evasive maneuver that will be performed after ejection of the LM/CSM from the S-IVB/IU/SLA.

TABLE 3

APOLLO 11 WEIGHT SUMMARY
(Weight in Pounds)

STAGE/MODULE	INERT WEIGHT	TOTAL EXPENDABLES	TOTAL WEIGHT	FINAL SEPARATION WEIGHT
S-IC Stage	288,750	4,739,320	5,028,070	363,425
S-IC/S-II Interstage	11,465	———	11,465	———
S-II Stage	79,920	980,510	1,060,430	94,140
S-II/S-IVB Interstage	8,080	———	8,080	———
S-IVB Stage	25,000	237,155	262,155	28,275
Instrument Unit	4,305	———	4,305	———
Launch Vehicle at Ignition		6,374,505		
Spacecraft-LM Adapter	4,045	———	4,045	———
Lunar Module	9,520	23,680	33,200	*33,635
Service Module	10,555	40,605	51,160	11,280
Command Module	12,250	———	12,250	11,020 (Landing)
Launch Escape System	8,910	———	8,910	———
Spacecraft At Ignition		109,565		
Space Vehicle at Ignition		6,484,070		
S-IC Thrust Buildup		(-)85,845		
Space Vehicle at Liftoff		6,398,325		
Space Vehicle at Orbit Insertion		292,865		

*CSM/LM Separation

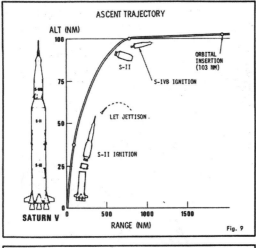

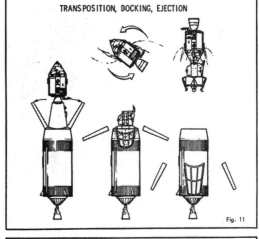

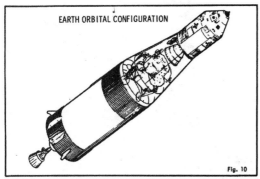

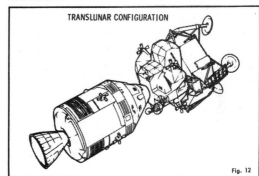

Translunar Coast

Within 2.5 hours after TLI, the CSM will be separated from the remainder of the vehicle and will transpose, dock with the LM, and initiate ejection of the CSM/LM from the SLA/IU/S-IVB as shown in Figure 11. A pitchdown maneuver of a prescribed magnitude for this transposition, docking, and ejection (TD&E) phase is designed to place the sun over the shoulders of the crew, avoiding CSM shadow on the docking interface. The pitch maneuver also provides continuous tracking and communications during the inertial attitude hold-during TD&E.

At approximately 1 hour 45 minutes after TLI, a spacecraft evasive maneuver will be performed using the SPS to decrease the probability of S-IVB recontact, to avoid ice particles expected to be expelled by the S-IVB during LOX dump, and to provide an early SPS confidence burn. This SPS burn will be performed in a direction and of a duration and magnitude that will compensate for the TLI bias mentioned before. The evasive maneuver will place the docked spacecraft, as shown in Figure 12, on a free return circumlunar trajectory. A free return to earth will be possible if the insertion into lunar parking orbit cannot be accomplished.

Approximately 2 hours after TLI, the residual propellants in the S-IVB are dumped to perform a retrograde maneuver. This "slingshot" maneuver reduces the probability of S-IVB recontact with the spacecraft and results in a trajectory that will take the S-IVB behind the trailing edge of the moon into solar orbit, thereby avoiding both lunar impact and earth impact.

Passive thermal control attitude will be maintained throughout most of the translunar coast period. Four midcourse correction maneuvers are planned and will be performed only if required. They are scheduled to occur at approximately TLI plus 9 hours, TLI plus 24 hours, lunar orbit insertion (LOI) minus 22 hours, and LOI minus 5 hours. These corrections will use the Manned Space Flight Network (MSFN) for navigation. The translunar coast phase will span approximately 73 hours.

Lunar Orbit Insertion

LOI will be performed in two separate maneuvers using the SPS of the CSM as shown in Figure 13. The first maneuver, LOI-1, will be initiated after the spacecraft has passed behind the moon and crosses the imaginary line through the centers of the earth and moon at approximately 80 NM above the lunar surface. The SPS burn is a retrograde maneuver that will place the spacecraft into an elliptical orbit that is approximately 60 x 170 NM. After two revolutions in the 60 x 170 NM orbit and a navigation update, a second SPS retrograde burn (LOI-2) will be made as the spacecraft crosses the antipode behind the moon to place the spacecraft in an elliptical orbit approximately 55 x 65 NM. This orbit will become circularized at 60 NM by the time of LM rendezvous due to the effect of variations in the lunar gravitational potential on the spacecraft as it orbits the moon.

LUNAR ORBIT INSERTION

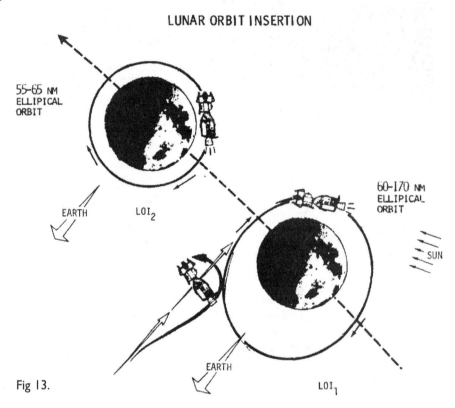

Fig 13.

CSM/LM Coast to LM Powered Descent

After LOI-2, some housekeeping will be accomplished in both the CSM and the LM. Subsequently, a simultaneous rest and eat period of approximately 10 hours will be provided for the three astronauts prior to checkout of the LM. Then the Commander (CDR) and Lunar Module Pilot (LMP) will enter the LM, perform a thorough check of all systems, and undock from the CSM. During the 13th revolution after LOI-2 and approximately 2.5 hours before landing, the LM and CSM will undock in preparation for descent. The undocking is a physical unlatching of a springloaded mechanism that imparts a relative velocity of approximately 0.5 feet per second (fps) between the vehicles. Station-keeping is initiated at a distance of 40 feet, and the LM is rotated about its yaw axis for CM Pilot observation of the deployed landing gear. Approximately one-half hour after undocking, the SM Reaction Control System (RCS) will be used to perform a separation maneuver of approximately 2.5 fps directed radially downward toward the center of the moon. This maneuver increases the LM/CSM separation distance to approximately 2.2 NM at descent orbit insertion (DOI). The DOI maneuver will be performed by a LM DPS retrograde burn, as shown in Figure 14, one-half revolution after LM/CSM separation. This maneuver places the LM in an elliptical orbit that is approximately 60 NM by 50,000 feet. The descent orbit events are shown in Figure 15.

DESCENT ORBIT INSERTION

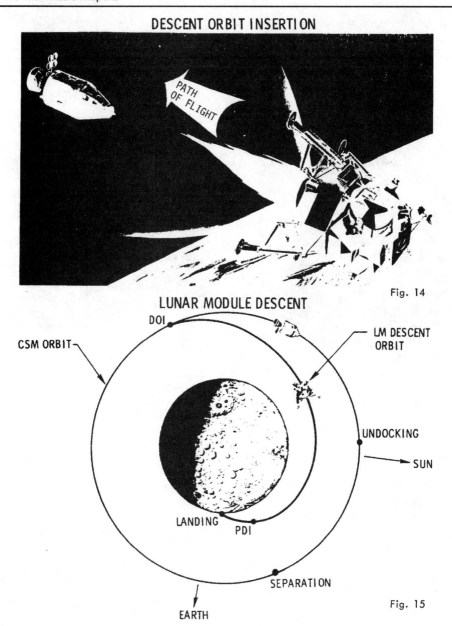

Fig. 14

LUNAR MODULE DESCENT

Fig. 15

Lunar Module Powered Descent

The LM powered descent maneuver will be initiated at the 50,000-foot altitude point of the descent orbit and approximately 14° prior to the landing site. This maneuver will consist of a braking phase, an approach phase, and a landing phase. The braking phase will use maximum thrust from the DPS for most of this phase to reduce the LM's orbital velocity. The LM will be rotated to a windows-up attitude at an altitude of 45,000 feet. The use of the landing radar can begin at an altitude of about 39,000 feet, as depicted in Figure 16. The approach phase, as shown in Figure 17, will begin at approximately 7600 feet (high gate) from the lunar surface. Vehicle attitudes during this phase will permit crew visibility of the landing area through the forward window. The crew can redesignate to an improved lunar surface area in the event the targeted landing point appears excessively rough. The landing phase will begin at an altitude of 500 feet (low gate) and has been designed to provide continued visual assessment of the landing site. The crew will take control of the spacecraft attitude and make minor adjustments as required in the rate of descent during this period.

The vertical descent portion of the landing phase will start at an altitude of 125 feet and continue at a rate of 3 fps until the probes on the foot pads of the LM contact the lunar surface. The CDR will cut off the descent engine within 1 second after the probes, which extend 68 inches beyond the LM footpad, contact the lunar surface although the descent engine can be left on until the footpads contact the lunar surface. The lunar surface contact sequence is shown in Figure 18.

LANDING RADAR-ANTENNA BEAM CONFIGURATION

 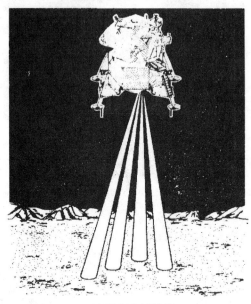

APPROACH PHASE   LANDING PHASE

Fig. 16

Lunar Surface Activities

Immediately after landing, the LM will be checked to assess its launch capability. After the postlanding checks and prior to preparation for extravehicular activity (EVA), there will be a 4-hour rest period, with eat periods before and after. A timeline for the lunar surface activity is shown in Figure 19. Each crewman will then don a "backpack" consisting of a Portable Life Support System (PLSS) and an Oxygen Purge System (OPS). The LM Environmental Control System (ECS) and the Extravehicular Mobility Unit (EMU) will be checked out, and the LM will be depressurized to allow the CDR to egress to the lunar surface. As the CDR begins to descend the LM ladder, he will pull a "D" ring which will lower the Modularized Equipment Stowage Assembly (MESA). This allows the TV camera mounted on the MESA access panel to record his descent to the lunar surface. The LMP will remain inside the LM Ascent Stage during the early part of the EVA to monitor the CDR's surface activity (including photography through the LM window) and the LM systems in the depressurized state.

Commander Environmental Familiarization

Once on the surface, the CDR will move-slowly from the footpad to check his balance and determine his ability to continue with the EVA - the ability to move and to see or, specifically, to perform the surface operations within the constraints of the EMU and the lunar environment. Although a more thorough evaluation and documentation of a crewman's capabilities will occur later in the timeline, this initial familiarization will assure the CDR that he and the LMP are capable of accomplishing the assigned EVA tasks. A brief check of the LM status will be made to extend the CDR's environment familiarization and, at the same time, provide an important contribution to the postflight assessment of the LM landing should a full or nominal LM inspection not be accomplished later.

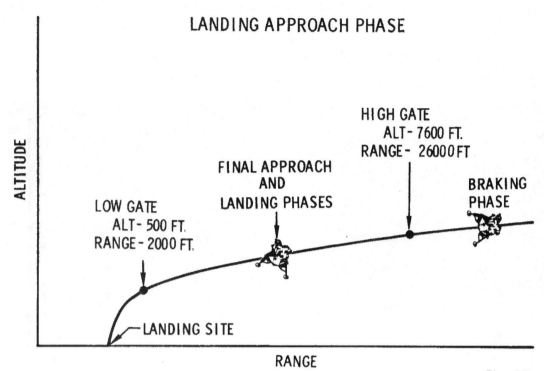

LANDING APPROACH PHASE

Fig. 17

LUNAR CONTACT SEQUENCE

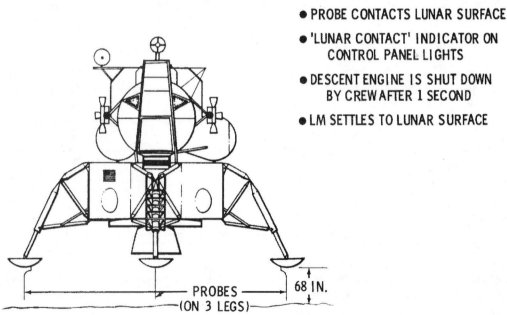

- PROBE CONTACTS LUNAR SURFACE
- 'LUNAR CONTACT' INDICATOR ON CONTROL PANEL LIGHTS
- DESCENT ENGINE IS SHUT DOWN BY CREW AFTER 1 SECOND
- LM SETTLES TO LUNAR SURFACE

Fig. 18

Contingency Sample Collection

A Contingency Sample of lunar surface material will be collected. This will assure the return of a small sample in a contingency situation where a crewman may remain on the surface for only a short period of time. One to four pounds of loose material will be collected in a sample container assembly which the CDR carries to the surface in his suit pocket. The sample will be collected near the LM ladder and the sample bag restowed

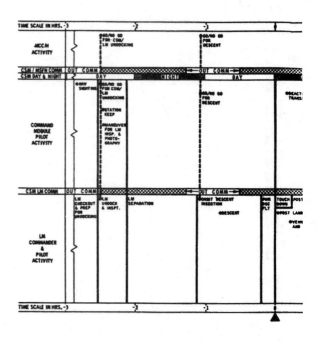

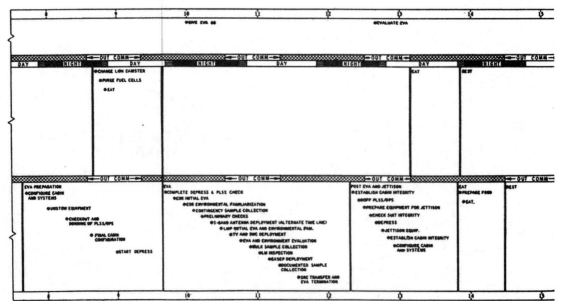

in the suit pocket to be carried into the Ascent Stage when the CDR ingresses at the end of the EVA. Figure 20 shows the relative location of the Contingency Sample collection and the other lunar surface activities.

S-Band Erectable Antenna Deployment

In the event that adequate margins do not exist with the steerable antenna for the entire communications spectrum (including television) during the EVA period, the S-band erectable antenna may be deployed to improve these margins. This would require approximately 19 minutes and will probably reduce the time allocated to other EVA events.

LUNAR SURFACE ACTIVITY
TIMELINE FOR 22-HOUR STAY

M-932-69-11

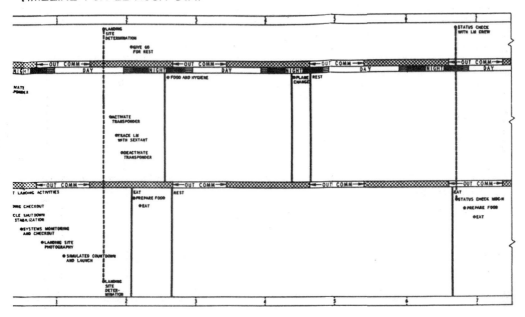

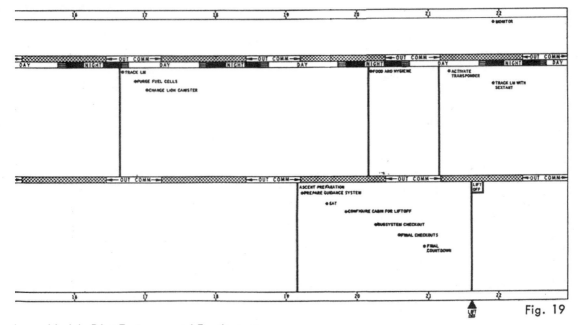

Fig. 19

Lunar Module Pilot Environmental Familiarization

After the CDR accomplishes the preliminary EVA task, the LMP will descend to the surface and spend a few minutes in the familiarization and evaluation of his capability or limitations to conduct further operations in the lunar environment.

Television Camera Deployment

The CDR, after photographing the LMP's egress and descent to the surface, will remove the TV camera from the Descent Stage MESA, obtain a panorama, and place the camera on its tripod in a position to view the subsequent surface EVA operations. The TV camera will remain in this position.

LUNAR SURFACE ACTIVITY

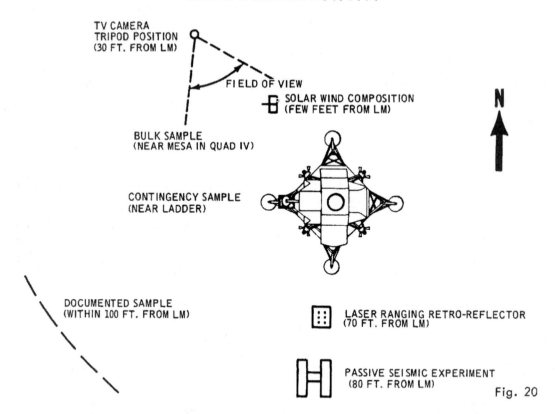

Fig. 20

Extravehicular Activity and Environmental Evaluation

The LMP will proceed to conduct the environmental evaluation. This involves a detailed investigation and documentation of a crewman's capability within the constraints of the EMU; the PLSS/EMU performance under varying conditions of sunlight, shadow, crewman activity or inactivity; and the characteristics of the lunar environment which influence operations on the surface.

Flag Deployment

Early in the LMP EVA period the astronauts will erect a 3 by 5-foot American flag. It will be on an 8-foot aluminum staff and a spring-like wire along its top edge will keep it unfurled in the airless environment of the moon. The event will be recorded on television and transmitted live to earth. The flag will be placed a sufficient distance from the LM to avoid damage by the ascent engine exhaust at lunar takeoff.

Bulk Sample Collection

The CDR will collect a Bulk Sample of lunar surface material. In the Bulk Sample collection at least 22 pounds, but as much as 50 pounds, of unsorted surface material and selected rock chunks will be placed in a special container, a lunar Sample Return Container (SRC), to provide a near vacuum environment for its return to the Lunar Receiving Laboratory (LRL). Apollo Lunar Handtools (ALHT), stowed in the MESA with the SRC, will be used to collect this large sample of loose lunar material from the surface near the MESA in Quad IV of the LM. Figure 21 shows the removal of tools stowed in the MESA. Figure 22 shows the preparation of a handtool for use. As each rock sample or scoop of loose material is collected, it will be placed into a large sample bag. Placing the sealed bag, rather than the loose material directly into the SRC prevents contamination and possible damage to the container seals.

Solar Wind Composition Experiment Deployment

The LMP will deploy the Solar Wind Composition (SWC) experiment. The SWC experiment consists of a panel of very thin aluminum foil rolled and assembled into a combination handling and deployment container. It is stowed in the MESA. Once the thermal blanket is removed from around the MESA equipment it is a simple task to remove the SWC, deploy the staff and the foil "window shade," and place it in direct sunlight where the foil will be exposed to the sun's rays, as shown in Figure 23. The SWC experiment is designed to entrap noble gas constituents of the solar wind, such as helium, neon, argon, krypton, and xenon. It is deployed early in the EVA period for maximum exposure time. At the conclusion of the EVA, the foil is rolled up, removed from the staff, and placed in a SRC. At the time the foil is recovered, the astronaut will push the staff into the lunar surface to determine, for postflight soil mechanics analysis, the depth of penetration.

Fig. 21   REMOVAL OF STOWED TOOLS FROM MESA

Lunar Module Inspection

The LMP will begin the LM inspection and will be joined by the CDR after the Bulk Samples have been collected. The purpose of the LM inspection is to visually check and photographically document the external condition of the LM landing on the lunar surface. The inspection data will be used to verify the LM as a safe and effective vehicle for lunar landings. The data will also be used to gain more knowledge of the lunar surface characteristics. In general the results of the inspection will serve to advance the equipment design and the understanding of the environment in which it operates. The crewmen will methodically inspect and report the status of all external parts and surfaces of the LM which are visible to them. The still color photographs will supplement their visual documentation for postflight engineering analysis and design verification. They will observe and photograph the RCS effects on the LM, the interactions of the surface and footpads, and the DPS effects on the surface as well as the general condition of all quadrants and landing struts.

PREPARATION OF HAND TOOL

Fig. 22

Early Apollo Scientific Experiments Package

When the crewmen reach the scientific equipment bay in Quad II, the LMP will open it and remove the Early Apollo Scientific Experiments Package (EASEP) using prerigged straps and pulleys as the CDR completes the LM inspection and photographically documents the LMP's activity. EASEP consists of two basic experiments: the Passive Seismic Experiment (PSE) and the Laser Ranging Retro-Reflector (LRRR). Both experiments are independent, self-contained packages weighing a total of about 170 pounds and occupying 12 cubic feet of space.

DEPLOYED SOLAR WIND COMPOSITION EXPERIMENT   Fig. 23

The PSE uses three long-period seismometers and one short-period vertical seismometer for measuring meteoroid impacts and moonquakes as well as to gather information on the moon's interior such as the existence of a core and mantle. The Passive Seismic Experiment Package (PSEP) has four basic subsystems; the structure/thermal subsystem provides shock, vibration, and thermal protection; the electrical power subsystem generates 34 to 46 watts by solar panel array; the data subsystem receives and decodes MSFN uplink commands and downlinks experiment data, handles power switching tasks; and the Passive Seismic Experiment subsystem measures lunar seismic activity with long-period and short-period seismometers which detect inertial mass displacement. Also included in this package are 15-watt radioisotope heaters to maintain the electronic package at a minimum of -60°F during the lunar night.

The LRRR experiment is a retro-reflector array with a folding support structure for aiming and aligning the array toward earth. The array is built of cubes of fused silica. Laser ranging beams from earth will be reflected back to their point of origin for precise measurement of earth-moon distances, center of moon's mass motion, lunar radius, earth geophysical information, and development of space communication technology.

Earth stations that will beam lasers to the LRRR include the McDonald Observatory at Fort Davis, Texas; the Lick Observatory in Mount Hamilton, California; and the Catalina Station of the University of Arizona. Scientists in other countries also plan to bounce laser beams off the LRRR.

In nominal deployment, as shown in Figures 24 through 26, the EASEP packages are removed individually from the storage receptacle and carried to the deployment site simultaneously. The crewmen will select a level site, nominally within ±15° of the LM -Y axis and at least 70 feet from the LM. The selection of the site is based on a compromise between a site which minimizes the effects of the LM ascent engine during liftoff, heat and contamination by dust and insulation debris (kapton) from the LM Descent Stage, and a convenient site near the scientific equipment bay.

Documented Sample Collection

After the astronauts deploy the EASEP, they will select, describe as necessary, and collect lunar samples, as shown in Figure 27, until they terminate the EVA. The Documented Sample will provide a more detailed and selective variety of lunar material than will be obtained from the Contingency and Bulk Samples. It will include a core sample collected with a drive tube provided in the Sample Return Container, a gas analysis sample collected by placing a representative sample of the lunar surface material in a special gas analysis container, lunar geologic samples, and descriptive photographic coverage of lunar topographic features.

Samples will be collected using tools stored in the MESA and will be documented by photographs. Samples will be placed individually in prenumbered bags and the bags placed in the Sample Return Container.

Television and Photographic Coverage

The primary purpose of the TV is to provide a supplemental real-time data source to assure or enhance the scientific and operational data return. It may be an aid in determining the exact LM location on the lunar surface, in evaluating the EMU and man's capabilities in the lunar environment, and in documenting the sample collections. The TV will be useful in providing continuous observation for time correlation of crew activity with telemetered data, voice comments, and photographic coverage.

Photography consists of both still and sequence coverage using the Hasselblad camera, the Maurer data acquisition camera, and the Apollo Lunar Surface Close Up Camera (ALSCC). The crewmen will use the Hasselblad extensively on the surface to document each major task which they accomplish. Additional photography, such as panoramas and scientific documentation, will supplement other data in the postflight analysis of the lunar environment and the astronauts' capabilities or limitations in conducting lunar surface operations. The ALSCC is a stereo camera and will be used for recording the fine textural details of the lunar surface material. The data acquisition camera (Sequence camera) view from the LM Ascent Stage window will provide almost continuous coverage of the surface activity. The LMP, who remains inside the Ascent Stage for the first few minutes of the EVA, will use the sequence camera to document the CDR's initial surface activities. Then, before he egresses, the LMP will position the camera for optimum surface coverage while both crewmen are on the surface. After the first crewman(LMP) ingresses he can use the sequence camera to provide coverage of the remaining surface activity.

EARLY APOLLO SCIENTIFIC EXPERIMENTS PACKAGE DEPLOYMENT

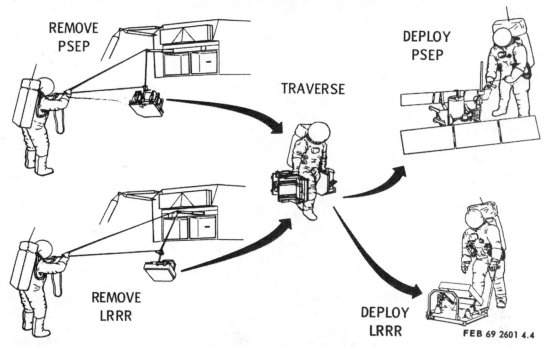

REMOVE PSEP

DEPLOY PSEP

TRAVERSE

REMOVE LRRR

DEPLOY LRRR

FEB 69 2601 4.4

PSEP- PASSIVE SEISMIC EXPERIMENTS PACKAGE
LRRR= LASER RANGING RETRO REFLECTOR

Fig. 24

Fig. 25   DEPLOYED PASSIVE SEISMIC EXPERIMENT

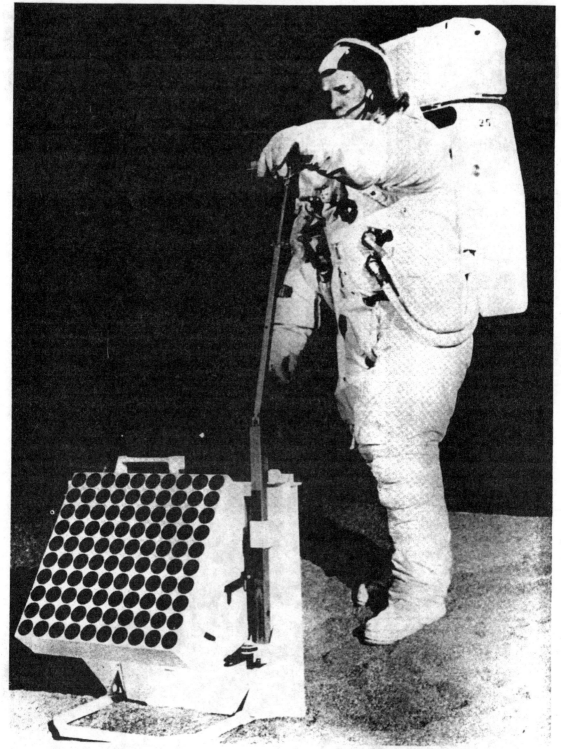

DEPLOYED LASER RANGING RETRO-REFLECTOR

Fig. 26

Extravehicular Activity Termination

The LMP will ingress before the SRC's are transferred to the LM. He will assist during the SRC transfer and will also make a LM systems check, change the sequence camera film magazine, and reposition the camera to cover the SRC transfer and the CDR's ladder ascent.

Fig. 27   DOCUMENTED SAMPLE COLLECTION

As each man begins his EVA termination he will clean the EMU. Although the crew will have a very limited capability to remove lunar material from their EMU's they will attempt to brush off any dust or particles from the portions of the suit which they can reach and from the boots on the footpad and ladder.

In the EVA termination there are two tasks that will require some increased effort. The first is the ascent from the footpad to the lowest ladder rung. In the unstroked position the vertical distance from the top of the footpad to the lowest ladder rung is 31 inches. In a nominal level landing this distance will be decreased only about 4 inches. Thus, unless the strut is stroked significantly the crewman is required to spring up using his legs and arms to best advantage to reach the bottom rung of the ladder from the footpad.

The second task will be the ingress or the crewmen's movement through the hatch opening to a standing position inside the LM. The hatch opening and the space inside the LM are small. Therefore, the crewmen must move slowly to prevent possible damage to their EMU's or to the exposed LM equipment.

After the crewmen enter the LM, they will jettison the equipment they no longer need. The items to be jettisoned are the used ECS canister and bracket, OPS brackets (adapters), and three armrests. The crewmen will then close the hatch and pressurize the LM. The EVA is considered to be terminated after the crewmen start this initial cabin pressurization. After the cabin pressure has stabilized, the crewmen will doff their PLSS's, connect to the LM ECS, and prepare to jettison more equipment they no longer need. The equipment, such as the PLSS's, lunar boots, and cameras, will be stowed in two containers. The LM will again be depressurized, the hatch opened, the containers jettisoned, and the cabin repressurized. Table 4 shows the loose equipment left on the lunar surface.

TABLE 4

LOOSE EQUIPMENT LEFT ON LUNAR SURFACE

During EVA

TV equipment
camera
tripod
handle/cable assembly
MESA bracket

Solar Wind Composition staff

Apollo Lunar Handtools
scoop
tongs
extension handle
hammer
gnomon

Equipment stowed in Sample Return Containers (outbound)
extra York mesh packing material
SWC bag (extra)
spring scale
unused small sample bags
two core tube bits
two SRC seal protectors
environmental sample
 containers O rings

Apollo Lunar Surface Close-up Camera (film cassette returned)

Hasselblad EL Data Camera (magazine returned)

EVA termination
Lunar equipment conveyor
ECS canister and bracket
OPS brackets
Three armrests

Post-EVA equipment jettison
Two Portable Life Support Systems
Left hand side stowage compartment (with equipment - such as lunar boots - inside)
One armrest

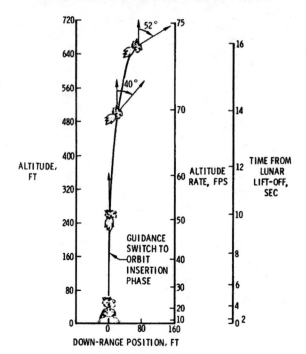

LUNAR MODULE VERTICAL RISE PHASE

Fig. 28

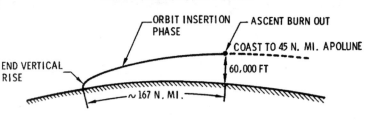

ORBIT INSERTION PHASE

TOTAL ASCENT:
BURN TIME = 7:15 MIN:SEC
ΔV REQUIRED = 6,056 FPS
PROPELLANT REQUIRED = 4,980 LBS

Fig. 29

Following the EVA and post-EVA activities, there will be another rest period of 4 hours 40 minutes duration, prior to preparation for liftoff.

Command/Service Module Plane Change

The CSM will perform a plane change of 0.18° approximately 2.25 revolutions after LM touchdown. This maneuver will permit a nominally coplanar rendezvous by the LM.

Lunar Module Ascent to Docking

After completion of crew rest and ascent preparations, the LM Ascent Propulsion System (APS) and the LM RCS will be used for powered ascent, rendezvous, and docking with the CSM.

Powered ascent will be performed in two phases during a single continuous burn of the ascent engine. The first phase will be a vertical rise, as shown in Figure 28, required for the Ascent Stage to clear the lunar terrain. The second will be an orbital insertion maneuver which will place the LM in an orbit approximately 9 x 45 NM. Figure 29 shows the LM ascent through orbit insertion. Figure 30 shows the complete rendezvous maneuver sequence and the coverage capability of the rendezvous radar (RR) and the MSFN tracking. After insertion into orbit, the LM will compute and execute the coelliptic rendezvous sequence which nominally consists of four major maneuvers: concentric sequence initiation (CSI), constant delta height (CDH), terminal phase initiation (TPI), and terminal phase finalization (TPF). The CSI maneuver will be performed to establish the proper phasing conditions at CDH so that, after CDH is performed, TPI will occur at the desired time and elevation angle. CSI will nominally circularize the LM orbit 15 NM below that of the CSM. CSI is a posigrade maneuver that is scheduled to occur approximately at apolune. CDH nominally would be a small radial burn to make the LM orbit coelliptic with the orbit of the CSM. The CDH maneuver would be zero if both the CSM and LM orbits are perfectly circular at the time of CDH. The LM will maintain RR track attitude after CDH and continue to track the CSM. Meanwhile the CSM will maintain sextant/VHF ranging tracking of the LM. The TPI maneuver will be performed with the LM RCS thrusters approximately 38 minutes after CDH. Two midcourse corrections (MCC-1 and MCC-2) are scheduled between TPI and TPF, but are nominally zero. TPF braking will begin approximately 42 minutes after TPI and end with docking to complete approximately 3.5 hours of rendezvous activities. One lunar revolution, recently added to the flight plan, will allow LM housekeeping activities primarily associated with back contamination control procedures. Afterward, the LM crewmen will transfer to the CSM with the lunar samples and exposed film.

RENDEZVOUS MANEUVERS/RADAR COVERAGE

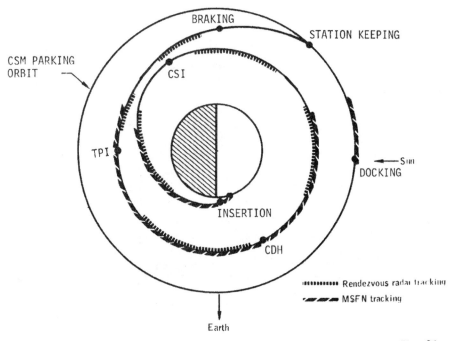

Fig. 30

Fig. 31   L U N A R   A C T I V

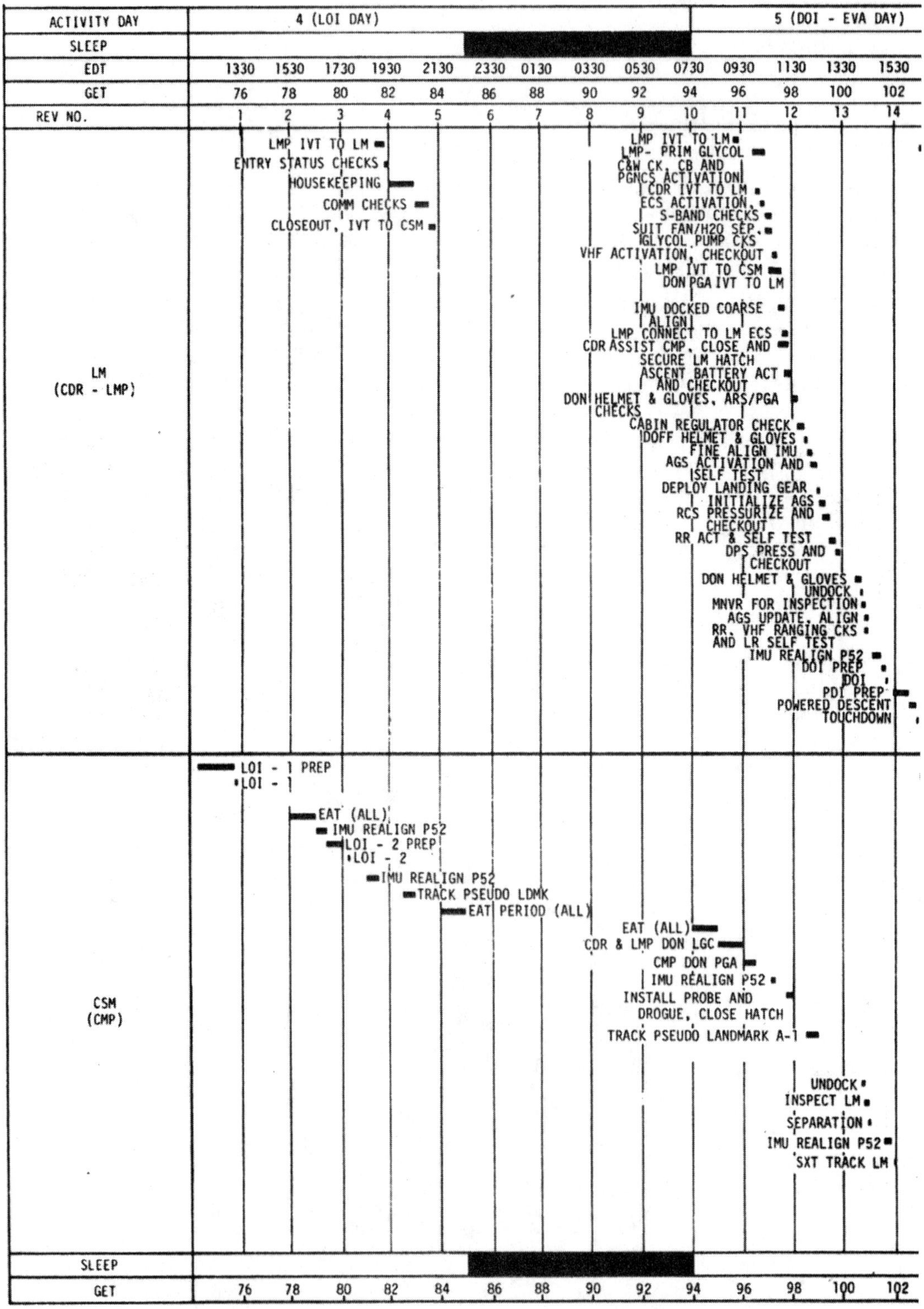

ACTIVITY DAY	4 (LOI DAY)										5 (DOI - EVA DAY)			
SLEEP						▮▮▮▮▮								
EDT	1330	1530	1730	1930	2130	2330	0130	0330	0530	0730	0930	1130	1330	1530
GET	76	78	80	82	84	86	88	90	92	94	96	98	100	102
REV NO.	1	2	3	4	5	6	7	8	9	10	11	12	13	14

LM (CDR - LMP)

- LMP IVT TO LM ▪
- ENTRY STATUS CHECKS ▬
- HOUSEKEEPING ▬
- COMM CHECKS ▬
- CLOSEOUT, IVT TO CSM ▪
- LMP IVT TO LM ▪
- LMP- PRIM GLYCOL ▬
- C&W CK, CB AND
- PGNCS ACTIVATION ▪
- CDR IVT TO LM ▪
- ECS ACTIVATION ▪
- S-BAND CHECKS ▪
- SUIT FAN/H2O SEP ▪
- GLYCOL PUMP CKS ▪
- VHF ACTIVATION, CHECKOUT ▪
- LMP IVT TO CSM ▬
- DON PGA IVT TO LM ▪
- IMU DOCKED COARSE ALIGN ▪
- LMP CONNECT TO LM ECS ▪
- CDR ASSIST CMP, CLOSE AND SECURE LM HATCH ▪
- ASCENT BATTERY ACT AND CHECKOUT ▪
- DON HELMET & GLOVES, ARS/PGA CHECKS ▪
- CABIN REGULATOR CHECK ▪
- DOFF HELMET & GLOVES ▪
- FINE ALIGN IMU ▪
- AGS ACTIVATION AND SELF TEST ▪
- DEPLOY LANDING GEAR ▪
- INITIALIZE AGS ▪
- RCS PRESSURIZE AND CHECKOUT ▪
- RR ACT & SELF TEST ▪
- DPS PRESS AND CHECKOUT ▪
- DON HELMET & GLOVES ▪
- UNDOCK ▪
- MNVR FOR INSPECTION ▪
- AGS UPDATE, ALIGN ▪
- RR, VHF RANGING CKS AND LR SELF TEST ▪
- IMU REALIGN P52 ▪
- DOI PREP ▪
- DOI ▪
- PDI PREP ▪
- POWERED DESCENT ▬
- TOUCHDOWN ▪

CSM (CMP)

- ▬ LOI - 1 PREP
- ▪ LOI - 1
- ▬ EAT (ALL)
- ▬ IMU REALIGN P52
- ▬ LOI - 2 PREP
- ▪ LOI - 2
- ▬ IMU REALIGN P52
- ▬ TRACK PSEUDO LDMK
- ▬ EAT PERIOD (ALL)
- EAT (ALL) ▬
- CDR & LMP DON LGC ▬
- CMP DON PGA ▪
- IMU REALIGN P52 ▪
- INSTALL PROBE AND DROGUE, CLOSE HATCH ▪
- TRACK PSEUDO LANDMARK A-1 ▬
- UNDOCK ▪
- INSPECT LM ▪
- SEPARATION ▪
- IMU REALIGN P52 ▬
- SXT TRACK LM ▪

SLEEP						▮▮▮▮▮								
GET	76	78	80	82	84	86	88	90	92	94	96	98	100	102

ITIES SUMMARY

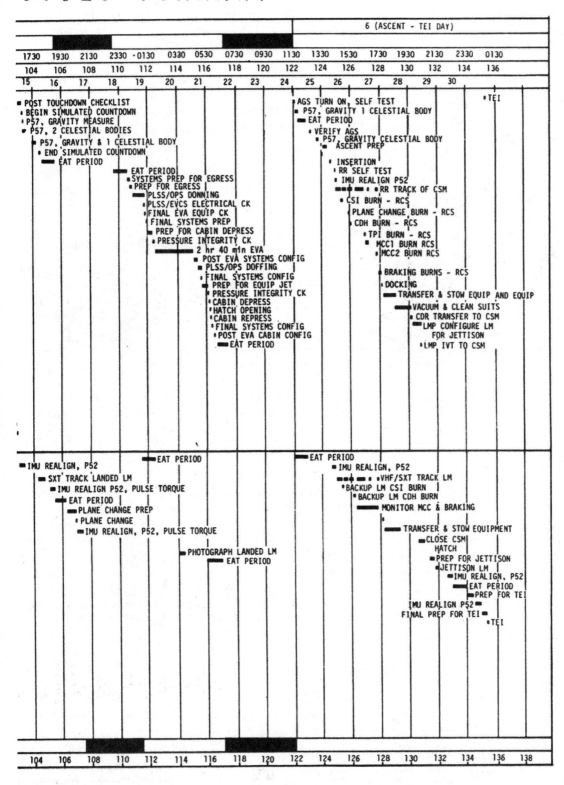

	6 (ASCENT - TEI DAY)

1730 1930 2130 2330 ·0130 0330 0530 0730 0930 1130 1330 1530 1730 1930 2130 2330 0130

104 106 108 110 112 114 116 118 120 122 124 126 128 130 132 134 136

15 16 17 18 19 20 21 22 23 24 25 26 27 28 29 30

- POST TOUCHDOWN CHECKLIST
- BEGIN SIMULATED COUNTDOWN
- P57, GRAVITY MEASURE
- P57, 2 CELESTIAL BODIES
  - P57, GRAVITY & 1 CELESTIAL BODY
  - END SIMULATED COUNTDOWN
    - EAT PERIOD

- EAT PERIOD
  - SYSTEMS PREP FOR EGRESS
  - PREP FOR EGRESS
    - PLSS/OPS DONNING
    - PLSS/EVCS ELECTRICAL CK
    - FINAL EVA EQUIP CK
    - FINAL SYSTEMS PREP
      - PREP FOR CABIN DEPRESS
      - PRESSURE INTEGRITY CK
        - 2 hr 40 min EVA
          - POST EVA SYSTEMS CONFIG
          - PLSS/OPS DOFFING
          - FINAL SYSTEMS CONFIG
          - PREP FOR EQUIP JET
          - PRESSURE INTEGRITY CK
          - CABIN DEPRESS
          - HATCH OPENING
          - CABIN REPRESS
            - FINAL SYSTEMS CONFIG
            - POST EVA CABIN CONFIG
              - EAT PERIOD

- AGS TURN ON, SELF TEST
  - P57, GRAVITY 1 CELESTIAL BODY
  - EAT PERIOD
    - VERIFY AGS
    - P57, GRAVITY CELESTIAL BODY
      - ASCENT PREP
    - INSERTION
    - RR SELF TEST
    - IMU REALIGN P52
      - RR TRACK OF CSM
    - CSI BURN - RCS
      - PLANE CHANGE BURN - RCS
    - CDH BURN - RCS
      - TPI BURN - RCS
      - MCC1 BURN RCS
      - MCC2 BURN RCS

      - BRAKING BURNS - RCS
      - DOCKING
        - TRANSFER & STOW EQUIP AND EQUIP
        - VACUUM & CLEAN SUITS
        - CDR TRANSFER TO CSM
        - LMP CONFIGURE LM
          FOR JETTISON
        - LMP IVT TO CSM

- TEI

- IMU REALIGN, P52
  - SXT TRACK LANDED LM
    - IMU REALIGN P52, PULSE TORQUE
    - EAT PERIOD
      - PLANE CHANGE PREP
      - PLANE CHANGE
        - IMU REALIGN, P52, PULSE TORQUE

- EAT PERIOD

- EAT PERIOD
  - IMU REALIGN, P52
    - VHF/SXT TRACK LM
    - BACKUP LM CSI BURN
    - BACKUP LM CDH BURN
      - MONITOR MCC & BRAKING

      - TRANSFER & STOW EQUIPMENT
        - CLOSE CSM
          HATCH
        - PREP FOR JETTISON
        - JETTISON LM
          - IMU REALIGN, P52
            - EAT PERIOD
            - PREP FOR TEI

      IMU REALIGN P52
      FINAL PREP FOR TEI
        - TEI

- PHOTOGRAPH LANDED LM
  - EAT PERIOD

104 106 108 110 112 114 116 118 120 122 124 126 128 130 132 134 136 138

Lunar Module Jettison to Transearth Injection

Approximately 2 hours after hard docking, the CSM will jettison the LM and then separate from the LM by performing a 1-fps RCS maneuver. The crew will then eat, photograph targets of opportunity, and prepare for transearth injection (TEI). Figure 31 presents a summary of activities from lunar orbit insertion through transearth injection.

Transearth Injection

The burn will occur 59.5 hours after LOI-1 as the CSM crosses the antipode on the far side of the moon. The spacecraft configuration for transearth injection and transearth coast is shown in Figure 32.

Transearth Coast

During transearth coast, three midcourse correction (MCC) decision points have been defined, as shown in Figure 33. The maneuvers will be targeted for corridor control only and will be made at the following times if required:

TRANSEARTH CONFIGURATION

MCC-5 -	TEI plus 15 hours
MCC-6 -	Entry interface (EI) minus 15 hours
MCC-7 -	EI minus 3 hours.

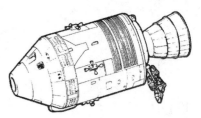

These corrections will utilize the MSFN for navigation. In the transearth phase there will be continuous communications coverage from the time the spacecraft appears from behind the moon until about 1 minute prior to entry. The constraints influencing the spacecraft attitude timeline are thermal control, communications, crew rest cycle, and preferred times of MCC's.

Fig. 32

The attitude profile for the transearth phase is complicated by more severe fuel slosh problems than for the other phases of the mission.

Entry Through Landing

Prior to atmospheric entry, the final MCC will be made and the CM will be separated from the SM using the SM RCS. The spacecraft will reach entry interface (EI) at 400,000 feet, as shown in Figure 34, with a velocity of 36,194 fps. The S-band communication blackout will begin 18 seconds later followed by C-band communication blackout 28 seconds from EI. The rate of heating will reach a maximum 1 minute 10 seconds after EI. The spacecraft will exit from C-band blackout 3 minutes 4 seconds after entry and from S-band blackout 3 minutes 30 seconds after entry. Drogue Parachute deployment will occur 8 minutes 19 seconds after entry at an altitude of 23,000 feet, followed by main parachute deployment at EI plus 9 minutes 7 seconds. Landing will occur approximately 14 minutes 2 seconds after and 1285 NM downrange from EI.

Landing will be in the Pacific Ocean at 172°W longitude, 11°N latitude and will occur approximately 8 days 3 hours after launch.

Postlanding Operations

Following landing, the recovery helicopter will drop swimmers who will install the flotation collar to the CM. A large, 7-man liferaft will be deployed and attached to the flotation collar. Biological Isolation Garments (BIG's) will be lowered into the raft, and one swimmer will don a BIG while the astronauts don BIG's inside the CM. Two other swimmers will move upwind of the CM on a second large raft. The postlanding ventilation fan will be turned off, the CM will be powered down, and the astronauts will egress to the raft. The swimmer will then decontaminate all garments, the hatch area, and the collar.

TRANSEARTH PHASE

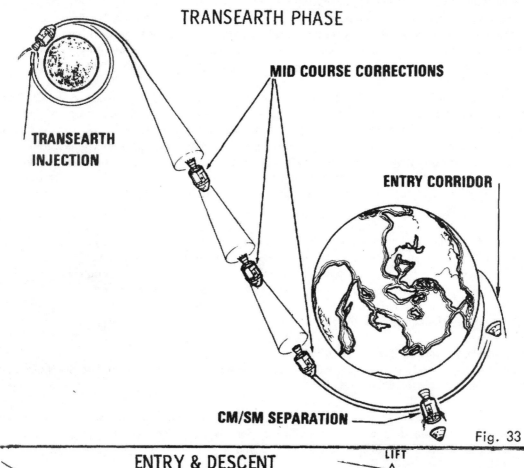

TRANSEARTH INJECTION

MID COURSE CORRECTIONS

ENTRY CORRIDOR

CM/SM SEPARATION

Fig. 33

ENTRY & DESCENT TO EARTH

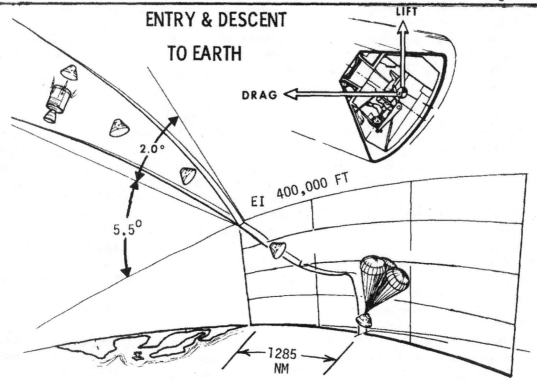

LIFT

DRAG

2.0°

5.5°

EI 400,000 FT

1285 NM

Fig. 34

The helicopter will recover the astronauts and the recovery physician riding in the helicopter will provide any required assistance. After landing on the recovery carrier, the helicopter will be towed to the hanger deck. The astronauts and the physician will then enter the Mobile Quarantine Facility (MQF). The flight crew, recovery physician and recovery technician will remain inside the MQF until it is delivered to the Lunar Receiving Laboratory (LRL) at the Manned Spacecraft Center (MSC) in Houston, Texas.

After flight crew pickup by the helicopter, the auxiliary recovery loop will be attached to the CM. The CM will be retrieved and placed in a dolly aboard the recovery ship. It will then be moved to the MQF and mated to the Transfer Tunnel. From inside the MQF/CM containment envelope, the MQF engineer will begin post-retrieval procedures (removal of lunar samples, data, equipment, etc.), passing the removed items through the decontamination lock. The CM will remain sealed during RCS deactivation and delivery to the LRL. The SRC, film, data, etc. will be flown to the nearest airport from the recovery ship for transport to MSC. The MQF and spacecraft will be offloaded from the ship at Pearl Harbor and then transported by air to the LRL.

In order to minimize the risk of contamination of the earth's biosphere by lunar material, quarantine measures will be enforced. The crew will be quarantined for approximately 21 days after liftoff from the lunar surface. In addition, the CM will be quarantined after landing. Termination of the CM quarantine period will be dependent on the results of the lunar sample analysis and observations of the crew.

BACK CONTAMINATION PROGRAM

The Apollo Back Contamination Program can be divided into three phases, as shown in Figure 35. The first phase covers the procedures which are followed by the crew while in flight to minimize the return of lunar surface contaminants in the Command Module.

The second phase includes spacecraft and crew recovery and the provisions for isolation and transport of the crew, spacecraft, and lunar samples to the Manned Spacecraft Center. The third phase encompasses the quarantine operations and preliminary sample analysis in the Lunar Receiving Laboratory (LRL).

A primary step in preventing back contamination is careful attention to spacecraft cleanliness following lunar surface operations. This includes use of special cleaning equipment, stowage provisions for lunar-exposed equipment, and crew procedures for proper "housekeeping."

LUNAR MODULE OPERATIONS

The Lunar Module (LM) has been designed with a bacterial filter system to prevent contamination of the lunar surface when the cabin atmosphere is released at the start of lunar exploration. Prior to reentering the LM after lunar surface exploration, the crewmen will brush any lunar surface dust or dirt from the space suit using the suit gloves. They will scrape their overboots on the LM footpad and while ascending the LM ladder dislodge any clinging particles by a kicking action. After entering the LM and pressurizing the cabin, the crew will doff their Portable Life Support System, Oxygen Purge System, lunar boots, EVA gloves, etc. The equipment to be jettisoned will be assembled and bagged to be subsequently left on the lunar surface. The lunar boots, likely the most contaminated items, will be placed in a bag as early as possible to minimize the spread of lunar particles. Following LM rendezvous and docking with the Command Module (CM), the CM tunnel will be pressurized and checks made to insure that an adequate pressurized seal has been made. During this period, the LM, space suits, and lunar surface equipment will be vacuumed. To accomplish this, one additional lunar orbit has been added to the mission.

The LM cabin atmosphere will be circulated through the Environmental Control System (ECS) suit circuit lithium hydroxide canister to filter particles from the atmosphere. A minimum of 5 hours of weightless operation and filtering will reduce the original airborne contamination to about 10^{-15} percent.

To prevent dust particles from being transferred from the LM atmosphere to the CM, a constant flow of 0.8 lb/hr oxygen will be initiated in the CM at the start of combined LM/CM operation. Oxygen will flow from the CM into the LM then overboard through the LM cabin relief valve or through spacecraft leakage. Since the flow of gas is always from the CM to the LM, diffusion and flow of dust contamination into the CM will be minimized. After this positive gas flow has been established from the CM, the tunnel hatch will be removed.

APOLLO BACK CONTAMINATION PROGRAM

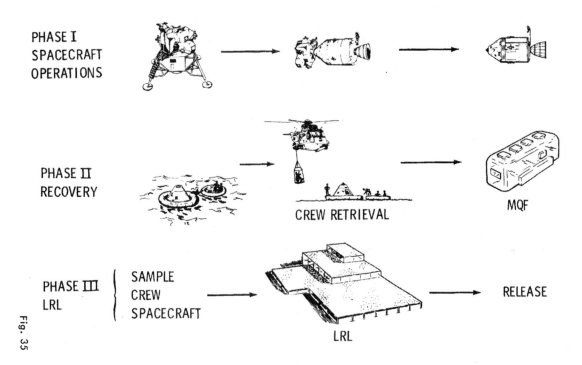

Fig. 35

The CM Pilot will transfer the lunar surface equipment stowage bags into the LM one at a time. The equipment transferred will then be bagged using the "buddy system" and transferred back into the CM where the equipment will be stowed. The only equipment that will not be bagged at this time are the crewmen's space suits and flight logs.

Following the transfer of the LM crew and equipment, the spacecraft will be separated and the three crewmen will start the return to earth. The separated LM contains the remainder of the lunar exposed equipment.

COMMAND MODULE OPERATIONS

Through the use of operational and housekeeping procedures the CM cabin will be purged of lunar surface and/or other particulate contamination prior to earth atmosphere entry. These procedures start while the LM is docked with the CM and continue through entry into the earth's atmosphere.

The LM crewmen will doff their space suits immediately upon separation of the LM and CM. The space suits will be stowed and will not be used again during the transearth phase unless an emergency occurs.

Specific periods for cleaning the spacecraft using the vacuum brush have been established. Visible liquids will be removed by the liquid dump system. Towels will be used by the crew to wipe surfaces clean of liquids and dirt particles. The three ECS suit hoses will be located at random positions around the spacecraft to insure positive ventilation, cabin atmosphere filtration, and avoid partitioning. During the transearth phase, the CM atmosphere will be continually filtered through the ECS lithium hydroxide canister. After about 63 hours operation, essentially none (10^{-90} percent) of the original contaminates will remain.

RECOVERY OPERATIONS

Following landing and the attachment of the flotation collar to the CM, the swimmer in a Biological Isolation Garment (BIG) will open the spacecraft hatch, pass three BIG's into the spacecraft, and close the hatch.

The crew will don the BIG's and then egress into the liferaft. The hatch will be closed immediately after egress. Tests have shown that the crew can don their BIG's in less than 5 minutes under ideal sea conditions. The spacecraft hatch will be open only for a few minutes. The spacecraft and crew will be decontaminated by the swimmer using a liquid agent. Crew retrieval will be accomplished by helicopter transport to the carrier. Subsequently, the crew will transfer to the Mobile Quarantine Facility. The spacecraft will be retrieved by the aircraft carrier.

BIOLOGICAL ISOLATION GARMENT

The BIG's will be donned in the CM just prior to egress and helicopter pickup and will be worn until the crew enters the Mobile Quarantine Facility aboard the primary recovery ship.

The suit is fabricated of a lightweight cloth fabric which completely covers the wearer and serves as a biological barrier. Built into the hood area is a face mask with a plastic visor, air inlet flapper valve, and an air outlet biological filter.

Two types of BIG's are used in the recovery operation. One is worn by the recovery swimmer. In this type garment, the inflow air (inspired) is filtered by a biological filter to preclude possible contamination of support personnel. The second type is worn by the astronauts. The inflow gas is not filtered, but the outflow gas (respired) is passed through a biological filter to preclude contamination of the air.

MOBILE QUARANTINE FACILITY

The Mobile Quarantine Facility (MQF) is equipped to house six people for a period up to 10 days. The interior is divided into three sections - lounge area, galley, and sleep/bath area. The facility is powered through several systems to interface with various ships, aircraft, and transportation vehicles. The shell is air and water tight. The principal method of assuring quarantine is to filter effluent air and provide a negative pressure differential for biological containment in the event of leaks.

Non-fecal liquids from the trailer are chemically treated and stored in special containers. Fecal wastes will be contained until after the quarantine period. Items are passed in or out of the MQF through a submersible transfer lock. A complete communications system is provided for intercom and external communications to land bases from ship or aircraft. Emergency alarms are provided for oxygen alerts while in transport by aircraft, for fire, loss of power, and loss of negative pressure.

Specially packaged and controlled meals will be passed into the facility where they will be prepared in a microwave oven. Medical equipment to complete immediate postlanding crew examination and tests are provided.

LUNAR RECEIVING LABORATORY

The final phase of the Back Contamination Program is completed in the Manned Spacecraft Center Lunar Receiving Laboratory (LRL). The crew and spacecraft are quarantined for a minimum of 21 days after lunar liftoff and are released based upon the completion of prescribed test requirements and results. During this time the CM will be disinfected. The lunar samples will be quarantined for a period of 50 to 80 days depending upon the results of extensive biological tests. The LRL serves four basic purposes:

* The quarantine of the lunar mission crew and spacecraft, the containment of lunar and lunar-exposed materials, and quarantine testing to search for adverse effects of lunar material upon terrestrial life.

* The preservation and protection of the lunar samples.

* The performance of time-critical investigation.

* The preliminary examination of returned samples to assist in an intelligent distribution of samples to principal investigators.

The LRL has a vacuum system with manually operated space gloves leading directly into a vacuum chamber at pressures of 10^{-7} torr (mm of mercury). It has a low-level counting facility with a background count an order of magnitude better than other known counters. Additionally, it is a facility that can handle cabinets to contain extremely hazardous pathogenic material.

The LRL covers 83,000 square feet of floor space and includes several distinct areas. These are the Crew Reception Area (CRA), Vacuum Laboratory, Sample Laboratories (Physical and Bioscience), and an administrative and support area. Special building systems are employed to maintain air flow into sample handling areas and the CRA to sterilize liquid waste and to incinerate contamination air from the primary containment systems.

The CRA provides biological containment for the flight crew and 12 support personnel. The nominal occupancy is about 14 days but the facility is designed and equipped to operate for considerably longer if necessary.

The biomedical laboratories provide for the required quarantine tests to determine the effect of lunar samples on terrestrial life. These tests are designed to provide data upon which to base the decision to release lunar material from quarantine.

Among the tests:

A. Germ-free mice will be exposed to lunar materials and observed continuously for 21 days for any abnormal changes. Periodically, groups will be sacrificed for pathologic observation.

B. Lunar material will be applied to 12 different culture media and maintained under several environmental conditions. The media will then be observed for bacterial or fungal growth. Detailed inventories of the microbial flora of the spacecraft and crew have been maintained so that any living material found in the sample testing can be compared against this list of potential contaminants taken to the moon by the crew or spacecraft.

C. Six types of human and animal tissue culture cells will be maintained in the laboratory and, together with embryonated eggs, will be exposed to the lunar material. Based on cellular and/or other changes, the presence of viral material can be established so that special tests can be conducted to identify and isolate the type of virus present.

D. Thirty-three species of plants and seedlings will be exposed to lunar material. Seed germination, growth of plant cells, or the health of seedlings will then be observed, and histological, microbiological, and biochemical techniques will be used to determine the cause of any suspected abnormality.

E. A number of lower animals will be exposed to lunar material. These specimens include fish, birds, oysters, shrimp, cockroaches, houseflies, planaria, paramecia, and euglena. If abnormalities are noted, further tests will be conducted to determine if the condition is transmissible from one group to another.

STERILIZATION AND RELEASE OF THE SPACECRAFT

Postflight testing and inspection of the spacecraft is presently limited to investigation of anomalies which happened during the flight. Generally, this entails some specific testing of the spacecraft and removal of certain components of systems for further analysis. The timing of postflight testing is important so that corrective action may be taken for subsequent flights.

The schedule calls for the spacecraft to be returned to port where a team will deactivate pyrotechnics, flush and drain fluid systems (except water). This operation will be confined to the exterior of the spacecraft. The spacecraft will then be flown to the LRL and placed in a special room for storage, sterilization, and postflight checkout.

APOLLO 11 NOMINAL MISSION EVENTS AND CONTINGENCY OPTIONS

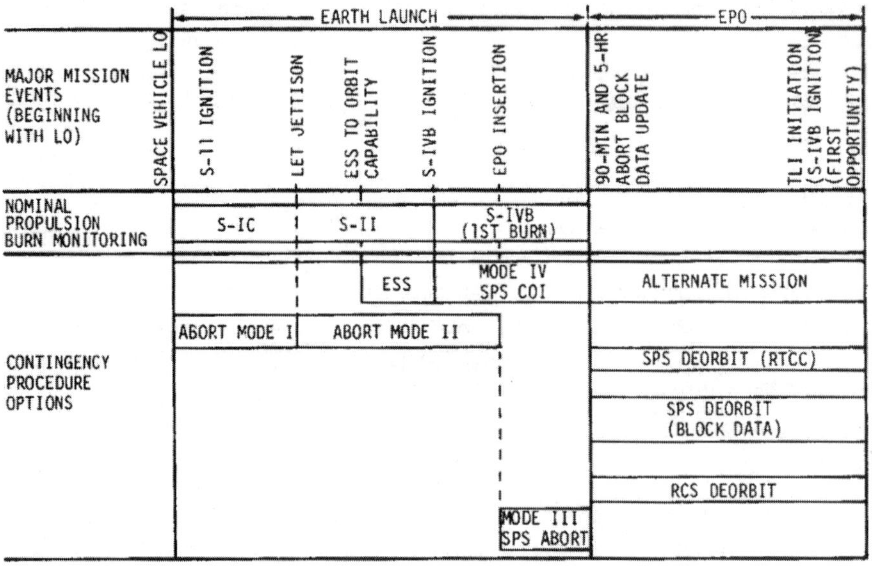

Fig. 36

CONTINGENCY OPERATIONS

GENERAL

If an anomaly occurs after liftoff that would prevent the space vehicle from following its nominal flight plan, an abort or an alternate mission will be initiated. Aborts will provide for an acceptable flight crew and CM recovery while alternate missions will attempt to maximize the accomplishment of mission objectives as well as providing for an acceptable flight crew and CM recovery. Figure 36 shows the Apollo 11 contingency options.

ABORTS

The following sections describe the abort procedures that may be used to return the CM to earth safely following emergencies that would prevent the space vehicle from following its normal flight plan. The abort descriptions are presented in the order of mission phase in which they could occur.

Launch

There are six launch abort modes. The first three would result in the termination of the launch sequence and a CM landing in the launch abort areas.

Mode I - The Mode I abort procedure is designed for safe recovery of the CM following an abort initiated between Launch Escape System arming and Launch Escape Tower jettison. The procedure would consist of the Launch Escape Tower pulling the CM off the launch vehicle and propelling it a safe distance downrange.

The resulting landing point would lie between the launch site and approximately 520 NM downrange.

Mode II - The Mode II abort could be performed from the time the Launch Escape Tower is jettisoned early during second-stage burn until the full-lift CM landing point reaches 3200 NM downrange. The procedure would consist of separating the CSM from the launch vehicle, separating the CM from the SM, and then letting the CM free fall to entry. The entry would be a full-lift, or maximum range trajectory, with a landing on the ground track between 440 and 3200 NM downrange.

Mode III - The Mode III abort procedure could be performed from the time the full-lift CM landing range reaches 3200 NM downrange until orbital insertion is achieved. The procedure would consist of separating the CSM from the launch vehicle and then, if necessary, performing a retrograde burn with the SPS so that the half-lift CM landing point is no farther than 3350 NM downrange. Since a half-lift entry would be flown, the CM landing point would be approximately 70° NM south of the ground track between 3000 and 3350 NM downrange.

Fig. 36(continued)

APOLLO 11 NOMINAL MISSION EVENTS AND CONTINGENCY OPTIONS (CONTINUED)

These three descriptions are based on aborts initiated from the nominal launch trajectory. Aborts from a dispersed trajectory will consist of the same procedures, but the times at which the various modes become possible and the resultant landing points may vary.

The following launch abort procedures are essentially alternate launch procedures and result in insertion of the spacecraft into a safe earth orbit. These procedures would be used in preference to Modes II and III above unless immediate return to earth is necessary during the launch phase.

Mode IV and Apogee Kick - The Mode IV abort procedure is an abort to earth parking orbit and could be performed any time after the SPS has the capability to insert the CSM into orbit. This capability begins approximately 8 minutes 30 seconds GET. The procedure consists of separating the CSM from the launch vehicle and, shortly afterwards, performing a posigrade SPS burn to insert the CSM into earth orbit. This means that any time during the S-IVB burn portion of the launch phase the CSM has the capability to insert itself into orbit if the S-IVB should fail. Apogee kick is a variation of the Mode IV abort wherein the SPS burn to orbit would be performed at, or near, the first spacecraft apogee. The main difference between the two is the time at which the posigrade SPS burn is performed.

S-IVB Early Staging - Under normal conditions, the S-IVB is inserted into orbit with enough fuel to perform

Fig. 36(continued)

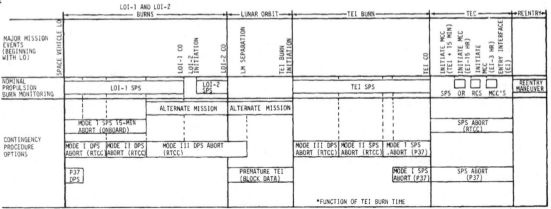

APOLLO 11 NOMINAL MISSION EVENTS AND CONTINGENCY OPTIONS
(CONTINUED)

the TLI maneuver. This capability can be used, if necessary, during the launch phase to insure that the spacecraft is inserted into a safe parking orbit. After approximately 6 minutes 30 seconds GET, the S-IVB has the capability to be staged early and achieve orbit. The CSM/LM could then remain in earth orbit to carry out an alternate mission, or, if necessary, return to the West Atlantic Ocean after one revolution.

S-IVB Early Staging to Mode IV - Should it become necessary to separate from a malfunctioning S-II stage, the S-IVB could impart sufficient velocity and altitude to the CSM to allow the SPS to be used to place the CSM into an acceptable earth orbit. The procedure is a combination of S-IVB early staging and Mode IV procedures. This means that at any time after 5 minutes 30 seconds GET the S-IVB/SPS combination may be utilized to boost the CSM into a safe earth orbit.

Earth Parking Orbit

Once the S-IVB/CSM is safely inserted into earth parking orbit, a return-to-earth abort would be performed by separating the CSM from the S-IVB and then utilizing the SPS for a retrograde burn to place the CM on on atmosphere-intersecting trajectory. After entry, the CM would be guided to a preselected target point, if available. This procedure would be similar to the deorbit and entry procedure performed on the Apollo 7 and Apollo 9 flights.

Translunar Injection

Ten-Minute Abort — There is only a remote possibility that an immediate return-to-earth will become necessary during the relatively short period of the TLI maneuver. However, if it should become necessary the S-IVB burn would be cut off early and the crew would initiate an onboard calculated retrograde SPS abort burn. The SPS burn would be performed approximately 10 minutes after TLI cutoff and would ensure a safe CM entry. The elapsed time from abort initiation to landing would vary from approximately 20 minutes to 5 hours, depending on the length of the TLI maneuver performed prior to S-IVB cutoff. For aborts initiated during the latter portion of TLI, a second SPS burn called a midcourse correction would be necessary to correct for dispersed entry conditions. Since this abort would be used only in extreme emergencies with respect to crew survival, the landing point would not be considered in executing the abort. No meaningful landing point predictions can be made because of the multiple variables involved including launch azimuth, location of TLI, the duration of the TLI burn prior to cutoff, and execution errors of the abort maneuvers.

Ninety-Minute Abort — A more probable situation than the previous case is that the TLI maneuver would be completed and then the crew would begin checking any malfunctions that may have been evident during the burn. If, after the check, it becomes apparent that it is necessary to return to earth, an abort would be initiated at approximately TLI cutoff plus 90 minutes. Unlike the previous procedure, this abort would be targeted to a preselected landing location called a recovery line. There are three recovery lines spaced around the earth as shown in Figure 37. This abort would be targeted to either the Mid-Pacific or the Atlantic Ocean recovery line. The abort maneuver would be a retrograde SPS burn followed by a midcourse correction, if necessary, to provide the proper CM entry conditions.

Translunar Coast

The CSM/LM will be in the translunar coast phase of the mission for approximately 3 days. The abort procedure during this time would be similar to the 90-minute abort. Abort information specifying a combination of SPS burn time and CSM attitude would be sent to the crew to be performed at a specific time. The longitude of the landing is determined by the time of abort and the abort trajectory. Therefore, fixed times of abort that will result in a landing on the Mid-Pacific recovery line will be selected during translunar coast. Because of the earth's rotation, a landing on the Mid-Pacific line can be accomplished only during one time interval for each 24-hour period. For this reason, a time critical situation may dictate targeting the abort to one of the other two recovery lines in order to minimize the elapsed time from abort to landing. The order of priority for the recovery lines is: (1) Mid-Pacific line, (2) Atlantic Ocean line, and (3) Indian Ocean line. Although the longitudes of the recovery lines are different, the latitude of landing will remain at approximately the latitude at which TLI occurred.

Fig. 37

RECOVERY LINES

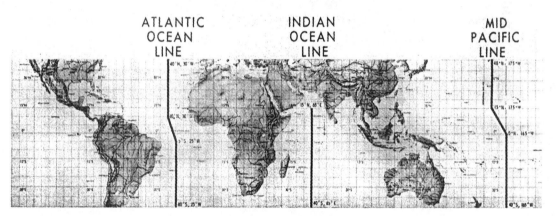

As the distance between the spacecraft and the moon decreases, the capability for return to earth increases. This continues until some time after the spacecraft reaches the moon's sphere of influence (basically, the point in the trajectory where the moon's influence on the spacecraft equals that of the earth) after which the return to earth becomes less for a circumlunar abort than for a direct return-to-earth abort.

Lunar Orbit Insertion

Should early termination of the LOI burn occur, the resulting abort procedure would be one of three modes classified according to length of burn before termination. Each abort mode would normally result in return of the CM to the Mid-Pacific recovery line. These modes are briefly discussed below.

Mode I - The Mode I procedure would be used for aborts following SPS cutoffs from ignition to approximately 1.5 minutes into the LOI burn. This procedure would consist of performing a posigrade DPS burn approximately 2 hours after cutoff to put the spacecraft back on a return-to-earth trajectory

Mode II - The Mode II procedure would be used for aborts following SPS shutdown during the interval approximately between LOI ignition plus 1.5 minutes and LOI ignition plus 3 minutes. This abort maneuver is performed in two stages. First, a DPS burn would be executed to reduce the lunar orbital period and to insure that the spacecraft does not impact on the lunar surface. After one orbit, a second DPS burn would place the spacecraft on a return-to-earth trajectory.

Mode III - The Mode III procedure would be used for aborts following shutdowns from approximately 3 minutes into the burn until nominal cutoff. After 3 minutes of LOI burn, the spacecraft will have been inserted into an acceptable lunar orbit. Therefore, the abort procedure would be to let the spacecraft go through one or two lunar revolutions prior to doing a posigrade DPS burn at pericynthion. This would place the spacecraft on a return-to-earth trajectory targeted to the Mid-Pacific recovery line.

Lunar Orbit

An abort from lunar orbit would be accomplished by performing the TEI burn early. Should an abort become necessary during the LM descent, ascent, or rendezvous phases of the mission, the LM would make the burns necessary to rendezvous with the CSM. If the LM were unable to complete the rendezvous, the CSM would, if possible, perform a rescue of the LM. In any case, the early TEI would normally target the CM to the Mid-Pacific recovery line.

Lunar Module Powered Descent

Aborts for the powered descent phase are controlled by the Primary Guidance and Navigation System (PGNS) abort program or the Abort Guidance System (AGS), depending on the operational status of the DPS and the PGNS. If both the PGNS and DPS are operational, the abort is initiated by pushing the "Abort" button. The DPS abort will continue under PGNS control until either orbit insertion or engine cutoff due to DPS failure or propellant depletion. If DPS cutoff occurs and the velocity-to-be-gained (VG) is less than 30 fps, the DPS will be staged manually and the RCS will be used to complete the orbit insertion of the LM. If VG is greater than 30 fps, the "Abort Stage" button is pushed. This stages the Descent Stage, and ignites the APS engine. The desired insertion orbit will then be obtained using the APS.

If the DPS has failed, the abort will be performed using the APS. As above, the procedure is to push the "Abort Stage" button.

If the PGNS is not operational, the abort is controlled by the AGS. For an operational DPS, the thrust level is controlled manually, and the steering is controlled by the AGS. If the DPS is not operational or becomes inoperative with a VG greater than 30 fps, the DPS will be staged manually, and the RCS will be used to insert the LM.

If both the PGNS and AGS have failed, a manual abort technique, using the horizon angle for a reference, will be used.

Lunar Stay

After LM touchdown if an early abort is required there are two preferred liftoff times. The first is actually a 15-minute span of time beginning at PDI (touchdown to touchdown plus 3 minutes). The second is at PDI plus 21.5 minutes (touchdown plus about 9.5 minutes). Both of these aborts will place the LM into a 9 x 30 NM orbit acceptable for LM-active rendezvous. Here again, an extra orbit and CSM dwell orbit are used to improve the rendezvous phasing and conditions in the former case and two revolutions are added in the latter case.

The above times may be adjusted somewhat in real-time to account for possible variations in the CSM orbit. Subsequently during the lunar stay, the preferred liftoff time is whenever the phasing is optimum for rendezvous. This occurs once each revolution shortly after the CM has passed over the site. The nominal rendezvous is performed with this phasing.

In the unlikely event of a catastrophic APS failure calling for an immediate liftoff, rendezvous following liftoff at any time could be performed within performance and time constraints. However, this contingency is considered highly unlikely and the rendezvous phasing is fairly poor during some periods. Due to the low probability of such an abort, highly developed operational plans for such are not being promulgated.

Aborts will proceed under the control of the PGNS, if operating, otherwise under control of AGS. A manual guidance scheme is being developed to provide backup in the event of both PGNS and AGS failure. This backup uses the Flight Director Attitude Indicator, if available, for attitude reference; otherwise, the horizon is used for reference.

Lunar Module Powered Ascent

Three types of aborts are available for the powered ascent phase. If the PGNS fails, the abort will require switching to the AGS. If the APS fails, the abort will be performed using the RCS for insertion provided the engine failure occurs within the RCS insertion capability. If both the PGNS and AGS fail, the abort will be performed by the crew using manual control.

Transearth Injection

The abort procedures for early cutoff of the SPS during the TEI burn are the inverse of the LOI abort procedures except that the abort would be performed by attempting to reignite the SPS. For SPS cutoff during the interval between TEI ignition and ignition plus 1.5 minutes, the Mode III LOI abort procedure would be used. For SPS cutoff between TEI ignition plus 1.5 minutes and TEI ignition plus 2 minutes, the Mode II LOI abort procedure would be used. If the SPS should be shut down from TEI ignition plus 2 minutes to nominal end of TEI, abort Mode I would be performed, except that the 2-hour coast period would be deleted.

Transearth Coast

From TEI until entry minus 24 hours, the only abort procedure that could be performed is to use the SPS or the SM RCS for a posigrade or retrograde burn that would respectively decrease or increase the transearth flight time and change the longitude of landing. After entry minus 24 hours, no further burns to change the landing point will be performed. This is to ensure that the CM maintains the desired entry velocity and flight path angle combination that will allow a safe entry.

Entry

If during entry, the Guidance, Navigation, and Control System (GNCS) fails, a guided entry to the end-of-mission target point cannot be flown. In this case, the crew would use their Entry Monitor System (EMS) to fly a 1285 NM range. The landing point would be approximately 39 NM uprange of the guided target point and 75 NM north of the ground track. If both the GNCS and EMS fail, a "constant g" (constant deceleration) entry would be flown. The landing point would be approximately 240 NM uprange of the guided target point and 75 NM north of the ground track.

ALTERNATE MISSION SUMMARY

The two general categories of alternate missions that can be performed during the Apollo 11 Mission are (1) earth orbital, and (2) lunar. Both of these categories have several variations which depend upon the nature of the anomaly causing the alternate mission and the resulting systems status of the LM and CSM. A brief description of these alternate missions is contained in the following paragraphs.

Earth Orbital Alternate Missions

Alternate 1 - CSM-Only Low Earth Orbit

Condition/Malfunction: LM not extracted, or S-IVB failed prior to 25,000-NM apogee, or SPS used to achieve earth orbit.

Perform: SPS LOI simulation (100 x 400-NM orbit), MCC's to approximate lunar timeline and for an approximate 10-day mission with landing in 150°W Pacific recovery area.
Alternate 2 - CSM-Only Semisynchronous

Condition/Malfunction: S-IVB fails during TLI with apogee >25,000 NM, LM cannot be extracted.

Perform: SPS phasing maneuver for LOI tracking, LOI simulation, SPS phasing maneuver to place perigee over Pacific recovery zone at later time, SPS semisynchronous orbit, and further MCC's to approximate lunar timeline.

Alternate 3 - CSM/LM Earth Orbit Combined Operations with SPS Deboost

Condition/Malfunction: TLI does not occur or TLI apogee <4000 NM, TD&E successful .

Perform: SPS maneuver to raise or lower apogee for orbit lifetime requirements if necessary, simulated LOI to raise or lower apogee to 400 NM, simulated DOI (in docked configuration), simulated PDI, SPS maneuver to circularize at 150 NM, a limited rendezvous (possibly CSM-active), and further SPS MCC's to complete lunar mission timeline.

Alternate 4 - CSM/LM Earth Orbit Combined Operations with DPS/SPS Deboost

Condition/Malfunction: S-IVB fails during TLI, SPS and DPS in combination can return CSM/LM to low earth orbit without sacrificing LM rescue (4000 NM <apogee < 10,000 NM).

Perform: SPS phasing maneuver, simulated DOI, PDI to lower apogee to about 4000 NM, SPS phasing (simulated MCC) maneuver to insure tracking for LOI, SPS maneuver to circularize at 150 NM, a limited rendezvous (possibly CSM-active), SPS maneuver to complete lunar mission, timeline, and achieve nominal 90 x 240-NM, end-of-mission orbit for an approximate 10-day mission with landing in 150°W Pacific recovery area.

Alternate 5 - CSM/LM Semisynchronous

Condition/Malfunction: SPS and DPS in combination cannot place CSM/LM in low earth orbit without sacrificing LM rescue, SPS propellant not sufficient for CSM/LM circumlunar mission.

Perform: SPS phasing maneuver (to place a later perigee over an MSFN site), SPS LOI (approximately semisynchronous), SPS phasing maneuver if necessary to adjust semisynchronous orbit, docked DPS DOI, docked DPS PDI simulation, SPS phasing to put perigee over or opposite recovery zone, SPS to semisynchronous orbit, and further MCC's to approximate lunar mission timeline.

Lunar Alternate Missions

Alternate To - DPS LOI

Condition/Malfunction: Non-nominal TLI such that: continuation of nominal mission, including CSM/LM LOI and TEI with SPS, is No-Go; but CSM/LM LOI Go with DPS LOI-1.

Perform: TD&E. SPS free-return CSM/LM, DPS LOI-1, and SPS LOI-2, after LOI-2, plane change for site coverage, photography and tracking of future landing sites, high-inclination orbit determination, SPS DOI with CSM/LM for three revolutions.

Alternate 1b - CSM Solo Lunar Orbit

Condition/Malfunction: Non-nominal TLI such that: CSM/LM LOI No-Go, CSM-only LOI Go.

Perform: TD&E, SPS free-return CSM/LM, LM testing during TLC and DPS staging, SPS plane changes in lunar orbit for additional site coverage, photography and tracking of future landing sites, high-inclination orbit determination, SPS to 60 x 8 NM orbit for three revolutions.

Alternate 1c - CSM/LM Flyby

Condition/Malfunction: Non-nominal TLI, such that: CSM/LM Flyby Go, CSM/LM LOI NoGo, CSM-only LOI No-Go.

Perform: TD&E, LM testing near pericynthion, docked DPS maneuver to raise pericynthion, DPS staging, and SPS for fast return.

Alternate 2 - CSM-Only Lunar Orbit

Condition/Malfunction: Failure to TD&E.

Perform: CSM-only lunar orbit mission, SPS plane change in lunar orbit for additional site coverage, photography and tracking of future landing sites, high inclination orbit determination, SPS to 60 x 8 NM orbit for three revolutions.

Alternate 3a - DPS TEI

Condition/Malfunction: LM No-Go for landing, but DPS Go for a burn.

Perform: SPS DOI to place CSM/LM in 60 x 8 NM orbit, three revolutions of tracking and photography, SPS circularization in 60 x 60 NM orbit, DPS TEI, SPS MCC for fast return.

Alternate 3b - DPS No-Go for Burn

Condition/Malfunction: LM No-Go for landing, and DPS No-Go for a burn.

Perform: CSM-only plane change for site coverage. Then follow same profile as Alternate 1b, above.

Alternate 4 - TEI With Docked Ascent Stage

Condition/Malfunction: CSM communications failure in lunar orbit.

Perform: TEI and keep LM as communication system. If DPS available, perform DPS TEI as in Alternate 3a. If Descent Stage jettisoned, perform SPS TEI with Ascent Stage attached.

CONFIGURATION DIFFERENCES

The space vehicle for Apollo 11 varies in its configuration from that flown on Apollo 10 and those to be flown on subsequent missions because of normal growth, planned changes, and experience gained on previous missions. Following is a list of the major configuration differences between AS-505 and AS-506.

SPACE VEHICLE	REMARKS
Command/Service Module (CSM-107)	
* Provided a short SPS main propellant sump tank.	To overcome potential delay in availability of scheduled tank.
* Changed insulation on hatch tunnel.	

SPACE VEHICLE	REMARKS

Lunar Module (Ascent Stage) (LM-5)

* Provided for first usage of EVA antenna

Required for lunar landing mission. (VHF).

* Incorporated Extravehicular Communication System (EVCS) into the PLSS.

Provides simultaneous and continuous telemetry from two extravehicular members, duplex voice communication between earth and one or both of the two extravehicular members, and uninterruptable voice communications between the crew members.

* Provided a Liquid Cooling Garment (LCG) heat removal subsystem.

Enhances mission success.

* Modified 22 critical stress corrosion fittings.

Enhances mission success.

Lunar Module (Descent Stage)

* Modified the base heat shield.

Reduces the lunar landing fire-to touchdown problem. Enhances mission success.

* Modified 11 critical stress corrosion fittings.

Enhances mission success.

* Added RCS plume deflectors for each of the lower four RCS thrusters.

To withstand increased firing time for RCS thrusters.

* Provided for first mission usage of erectable antenna (S-band).

Lunar landing mission requirement.

* Provided a modified gimbal drive actuator (polarizer and armature removed, added a new broke material and sleeve).

Enhances system performance.

Spacecraft-LM Adopter (SLA-14)

* (No significant differences.)

LAUNCH VEHICLE   REMARKS

Instrument Unit (S-IU-506)
* (No significant differences.)

S-IVB Stage (SA-506)
* (No significant differences.)

S-II Stage (S-II-506)

* Deleted research and development (R&D) instrumentation and retained operational instrumentation only.

Basic requirement.

S-IC Stage (SA-506)

* Retained operational instrumentation only.	Weight reduction of 5900 pounds results from deletion of R&D instrumentation.

MISSION SUPPORT

GENERAL

Mission support is provided by the Launch Control Center (LCC), the Mission Control Center (MCC), the Manned Space Flight Network (MSFN), and the recovery forces. The LCC is essentially concerned with prelaunch checkout, countdown, and with launching the SV, while MCC located at Houston, Texas, provides centralized mission control from liftoff through recovery. The MCC functions within the framework of a Communications, Command, and Telemetry System (CCATS); Real-Time Computer Complex (RTCC); Voice Communications System; Display/Control System; and a Mission Operations Control Room (MOCR) supported by Staff Support Rooms (SSR's). These systems allow the flight control personnel to remain in contact with the spacecraft, receive telemetry and operational data which can be processed by the CCATS and RTCC for verification of a safe mission, or compute alternatives. The MOCR is staffed with specialists in all aspects of the mission who provide the Mission Director and Flight Director with real-time evaluation of mission progress.

MANNED SPACE FLIGHT NETWORK

The MSFN is a worldwide communications and tracking network which is controlled by the MCC during Apollo missions. The network is composed of fixed stations (Figure 38) and is supplemented by mobile stations (Table 5) which are optimally located within a global band extending from approximately 40° south latitude to 40° north latitude. Station capabilities are summarized in Table 6. Figure 39 depicts communications during lunar surface operations.

The functions of these stations are to provide tracking, telemetry, updata, and voice communications both on an uplink to the spacecraft and on a downlink to the MCC. Connection between these many MSFN stations and the MCC is provided by NASA Communications Network (NASCOM). More detail on mission support is in the MOR Supplement.

TABLE 5

MSFN MOBILE FACILITIES

Ships	Location	Support
USNS VANGUARD	25°N 49°W	Insertion
USNS MERCURY	10°N 175.2°W	Injection
USNS REDSTONE	2.25°S 166. 8°E	Injection
USNS HUNTSVILLE	3. 0°N 154°E	Entry (tentative)

APOLLO RANGE INSTRUMENTATION AIRCRAFT

Eight Apollo Range Instrumentation Aircraft (ARIA) will be available to support the Apollo 11 Mission in the Pacific sector. The mission plan calls for ARIA support of translunar injection on revolution 2 or 3 and from entry (400,000-foot altitude) to recovery of the spacecraft and crew after landing.

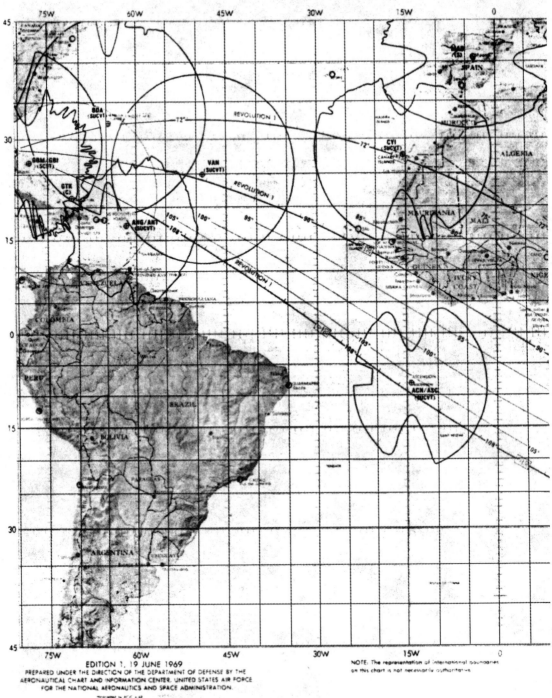

EDITION 1, 19 JUNE 1969
PREPARED UNDER THE DIRECTION OF THE DEPARTMENT OF DEFENSE BY THE
AERONAUTICAL CHART AND INFORMATION CENTER, UNITED STATES AIR FORCE
FOR THE NATIONAL AERONAUTICS AND SPACE ADMINISTRATION.

NOTE: The representation of international boundaries
on this chart is not necessarily authoritative.

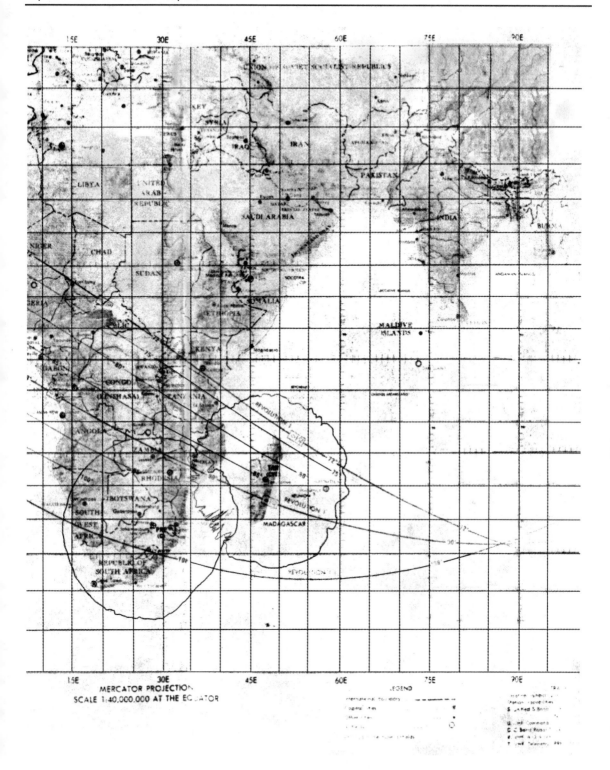

MERCATOR PROJECTION
SCALE 1:40,000,000 AT THE EQUATOR

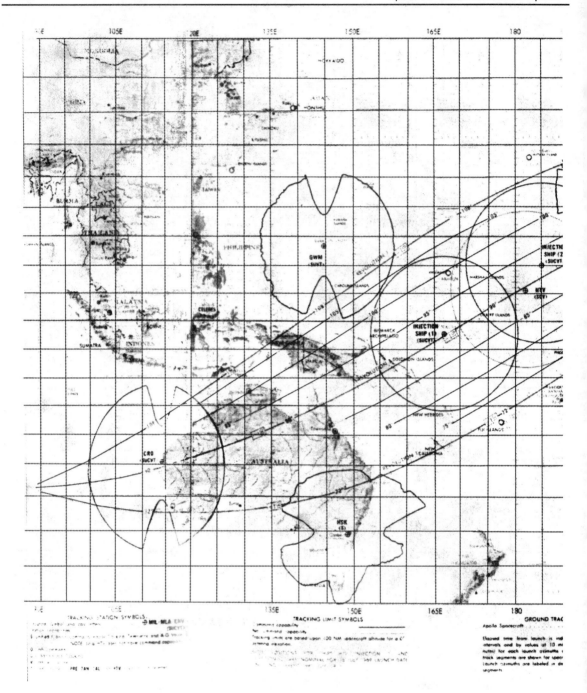

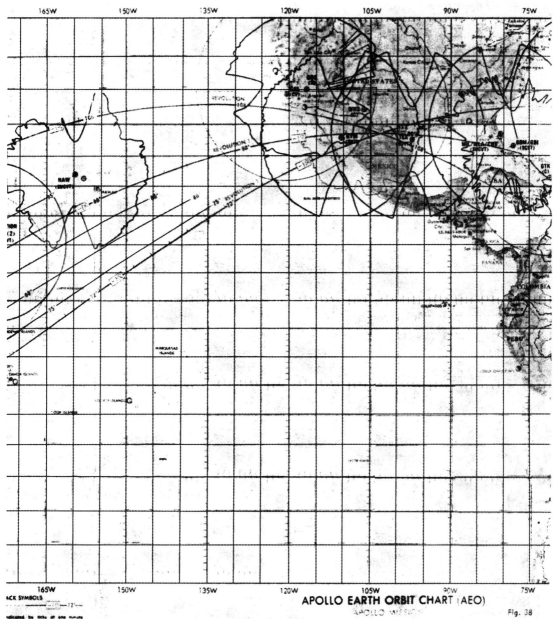

APOLLO EARTH ORBIT CHART (AEO)

APOLLO MISSION

Fig. 38

FOR JULY 1969 LAUNCH

TABLE 6
MSFN CONFIGURATION, APOLLO 11 MISSION

Station (B)	Type A	Type C	SPAN	USB (TV)	TV	USB (A/G Voice)	VHF (A/G Voice)	Cmd Destruct	Cmd Remoting	Cmd Processor	USB Update	Display	Bio-Med Remoting	Data Remoting	Data Processor	USB (TM)	VHF Links	USB (Tracking)	C-Band Radar
PAFB	x																		x
CNV	x	x						x											x
MLA	x																		x
CIF														x	x	x	x		
MIL				x		x	x		x	x	x			x	x	x	x	x	
GBI	x							x									x		
GBM	x			x		x	x		x	x	x			x	x	x	x	x	
GTK	x							x											
ANT	x							x											
ANG	x			x		x	x		x	x	x			x	x	x	x	x	
BDA	x	x		x		x	x		x	x	x			x	x	x	x	x	
CYI			x	x		x	x		x	x	x			x	x	x	x	x	
MAD				x*		x			x	x	x			x	x*	x		x	
MADX				x		x			x	x	x			x	x	x		x	
ASC		x																	
ACN				x		x			x	x	x			x	x	x		x	
PRE	x																		
TAN	x					x											x		
CRO	x		x	x		x	x		x	x	x			x	x	x	x	x	
HSK				x*		x			x	x	x			x	x	x		x	
HSKX				x		x			x	x	x			x	x	x		x	
GWM				x		x	x		x	x	x			x	x	x		x	
PARKES 210 ft.				x*		x											x		
HAW	x			x		x	x		x	x	x			x	x	x	x	x	
CAL	x						x												
GDS				x*		x			x	x	x			x	x	x		x	
GDSX				x		x			x	x	x			x	x	x		x	
GYM				x		x	x		x	x	x			x	x	x	x	x	
TEX	x			x*		x	x		x	x	x			x	x	x	x	x	
GDS 210 ft.				x*		x											x		
INS.	x	x				x	x		x	x	x			x	x	x	x	x	
INJ.	x	x		x		x	x		x	x	x			x	x	x	x	x	
INJ.	x	x		x		x	x		x	x	x			x	x	x	x	x	
RE-EN.	x	x				x	x					x		x		x	x	x	
A/RIA						x	x									x	x	x	

APOLLO LUNAR SURFACE COMMUNICATIONS

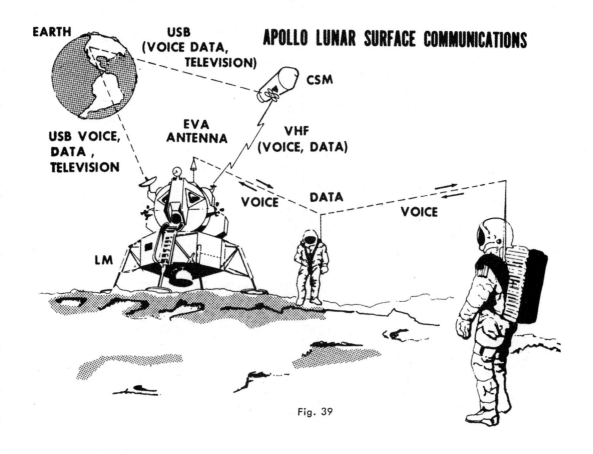

Fig. 39

RADAR COVERAGE DURING LUNAR ORBIT PERIODS FOR LAUNCH DATE OF JULY 16

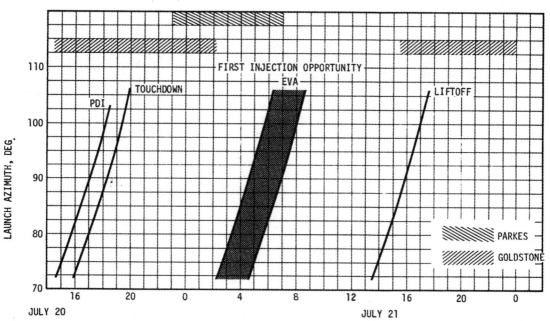

Fig. 40   EASTERN DAYLIGHT TIME, HR,

APOLLO 11 LAUNCH SITE AREA AND FORCE DEPLOYMENT

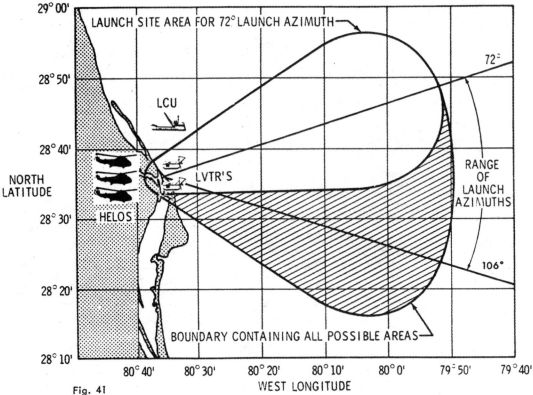

Fig. 41

RECOVERY SUPPORT PLAN

GENERAL

The Apollo 11 flight crew and Command Module (CM) will be recovered as soon as possible after landing, while observing the constraints required to maintain biological isolation of the flight crew, CM, and materials removed from the CM. After locating the CM, first consideration will be given to determining the condition of the astronauts and to providing first-level medical aid when required. The second consideration will be the recovery of the astronauts and CM. Retrieval of the CM main parachutes, apex cover, and drogue parachutes, in that order, is highly desirable if feasible and practical. Special clothing, procedures, and the Mobile Quarantine Facility (MQF) will be used to provide biological isolation of the astronauts and CM. The lunar sample rocks will also be isolated and returned to the Manned Spacecraft Center within 30 hours as specified by NASA.

The recovery forces will also be capable of salvaging portions of the space vehicle in case of a catastrophic failure in the vicinity of the launch site. Specific components to be recovered will be identified after the fact. After a normal launch, if items such as portions of the first stage of the launch vehicle or the Launch Escape System (LES) are found, they should be recovered if possible. If it appears that the items are too large or unsafe for retrieval, the Mission Control Center will be contacted for guidance before recovery is attempted.

LAUNCH PHASE

During the time between LES arming and parking orbit insertion, the recovery forces are required to provide support for landings that would follow a Mode I, II, or III launch abort

Launch Site Area

The launch site area includes all possible CM landing points which would occur following aborts initiated between LES arming and approximately 90 seconds GET. Figure 41 shows the launch site area and recovery force deployment. Recovery forces in the launch site area will be capable of meeting a maximum access time of 30 minutes to any point in the area. This support is required from the time the LES is armed until 90 seconds after liftoff. However, prior to LES arming, the launch site forces are required to be ready to provide assistance, if needed, to the Pad Egress Team, and, after T plus 90 seconds, they are required to be prepared to provide assistance to the launch abort area recovery forces. In addition to the 30-minute access time, the launch site recovery forces are required to have the capability to:

LAUNCH ABORT AREA AND FORCE DEPLOYMENT

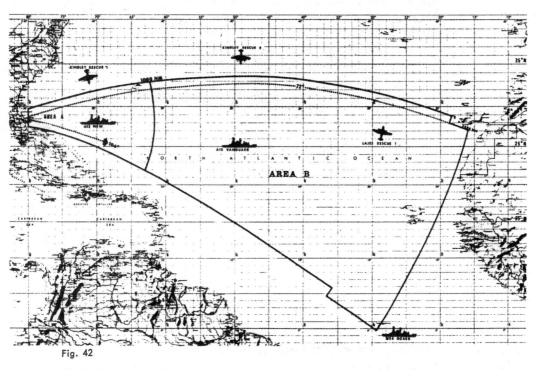

Fig. 42

a .   Provide firefighting units that are capable of containing hypergolic fuel fires.
b.   Upright the CM.
c.   Transport the flight crew from any point in the area to the Patrick AFB hospital.
d.   Transport the CM to a deactivation site.
e.   Provide debris location, mapping, and recording assistance for a salvage operation.

Launch Abort Area

The launch abort area is the area in which the CM would land following an abort initiated during the launch phase of flight, after approximately 90 seconds GET. The launch abort area shown in Figure 42 includes all possible CM landing points following a launch abort from any launch azimuth. The launch abort landing area is divided into two sectors: A and B. These sectors are used to differentiate the level of recovery support available in the area. Sector A is all the area in the launch abort area that is between 41 and 1000 nautical miles (NM) downrange of the launch site. Sector B is all the area in the launch abort area that is between 1000 and 3400 NM downrange of the launch site.

The primary responsibility of launch abort forces is to locate and recover the astronauts and retrieve the CM within the required access and retrieval times should a landing occur in the launch abort area. The forces required, their staging bases, and access times are listed in Table 7.

Two secondary recovery ships and three search and rescue aircraft will be positioned in the launch abort area as shown in Figure 42. Ship and aircraft stations in the launch abort area will sweep to the south each day during the launch window as the launch azimuth changes from 72° to 106°. Recovery ships and aircraft are positioned for optimum coverage of the 72° launch azimuth. Launch abort aircraft are required to provide a 4-hour access time to any launch azimuth. Retrieval time in Sector A will be 24 hours. Sector B will be considered as a contingency retrieval area; therefore, retrieval will be as soon as possible. HC-130 aircraft will be on station ten minutes prior to predicted landing time. Recovery forces providing immediate launch abort support will be released after translunar injection (TLI).

EARTH PARKING ORBIT PHASE

Earth parking orbit (EPO) secondary landing areas (SLA's) are configured to include target points and associated dispersion areas with low-speed entries from near-earth orbits. These areas are selected to provide recovery support at suitable time intervals throughout the EPO phase of the mission. The SLA is a 210-NM long by 80-NM wide dispersion ellipse oriented along the entry ground track and centered on the target point. For the Apollo 11 Mission, EPO SLA's will be required for four revolutions and are selected in two general locations called recovery zones. See Figure 43 for locations and Table 8 for access and retrieval times and forces required. If the TLI maneuver is not completed and a long duration earth orbital mission is flown, Zone 2, located in the East Atlantic Ocean, will also be activated. For landings during this phase of the mission, one HC-130 aircraft will be stationed 50 NM abeam of the target point.

TABLE 7

RECOVERY FORCE REQUIREMENTS

LAUNCH ABORT AREA

AREA OR ZONE	DESCRIPTION	RETRIEVAL TIME (HR) SHIP	ACCESS TIME (HR) A/C	SHIP			HC-130 AIRCRAFT		
				STA	POSITION	TYPE	STA	NO	POSITION
L A U N C H	Sector A: From launch site to 1000 NM downrange. 50 NM north of 072° azimuth and 50 NM south of 106° azimuth.	24	4	1	28°00'N 70°00'W	DD	A1	1	32°35'N 71°00'W
A B O R T	Sector B: From 1000 NM to 3400 NM downrange. 50 NM north of 072° azimuth and 50 NM south of 106° azimuth.	ASAP	4	2	25°00'N 49°00'W	AIS	B1	1	35°00'N 49°05'W
							C1	1	27°35'N 28°25'W

NOTE 1: Ship positions shown are for 072° azimuth launch on 16 July. As launch azimuth increases, ships will proceed south.
NOTE 2: Aircraft positions shown are for 072° azimuth launch on 16 July. As launch azimuth increases, aircraft will proceed south and maintain their relative position to the changing ground track.

The contingency landing area for this phase includes all the earth's surface between 34°N and 34°S latitude, except the launch abort, earth orbital, and deep space SLA's, and the end-of-mission planned landing area. The forces required, their staging bases, and access times are listed in Table 8.

DEEP SPACE PHASE

Deep space SLA's are designed to include the target point and dispersion area associated with a high-speed entry from space. These areas are selected to provide recovery support at suitable time intervals throughout the translunar, lunar orbit, and transearth phases of the mission. Deep space SLA's are located along or near ship-supported recovery lines which are spaced to provide varying return times as shown in Figure 44.

APOLLO 11 EARTH PARKING ORBIT RECOVERY ZONES

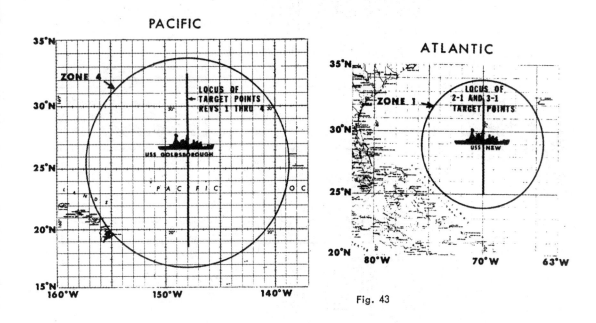

Fig. 43

For the Apollo 11 Mission, these areas are defined as the areas where a landing could occur following translunar coast aborts targeted to the Mid Pacific Line (MPL) (line 4), and any abort after TLI targeted to the Atlantic Ocean Line (AOL) (line 1). USS HORNET and two aircraft (HC-130's) are required to provide secondary landing area support for the MPL, and USS OZARK and two aircraft (HC-130's) are required for the AOL. The two ships will move along lines 1 and 4 to maintain the latitude of the moon's declination in the opposite hemisphere. Table 9 shows the approximate location of this point for each day's launch window. Actual positions required for each day will be published in the appropriate task force operations order.

The only time the ship's position is critical is during the first few hours after TLI. The minimum time between abort initiation and landing for these aborts will be approximately 11 hours for the AOL and 13 hours for the MPL. After these times, the return time becomes greater leaving sufficient time to position the ship at the CM target point. At entry minus 35 hours, if the CM is still targeted to the MPL, USS OZARK will be released.

Aborts made to the MPL or AOL after TLI require, within the high-speed entry footprint, an access time of 14 hours and retrieval time of 24 hours to any point in the area.

For deep space aborts to the MPL, one HC-130 aircraft will be stationed 200 NM uprange and 100 NM north of the ground track, one 200 NM north of ground track, and one abeam of the target point and 50 NM north of the ground track. Minimum alert posture for HC-130 aircraft is listed in Table 10. Table 11 shows the access and retrieval times and forces required to support the deep space SLA's.

The contingency landing area for the deep space phase of the mission is associated with very low probability of a CM landing and requires land-based recovery aircraft support only. For Apollo 11, the deep space contingency landing area is all the area in a band around the earth between 40°N and 15°S outside the primary and secondary landing areas. The forces required, their staging bases, and access times are shown in Table 8.

TABLE 8

RECOVERY FORCE REQUIREMENTS

EARTH ORBITAL PHASE

SECONDARY LANDING AREAS

RECOVERY ZONES	RETRIEVAL TIME (HR) SHIP	ACCESS TIME (HR) A/C	SHIP TYPE	SHIP POSITION	HC-130 AIRCRAFT NO	HC-130 AIRCRAFT STAGING BASES
1. 300 NM radius of 29°00'N, 70°00'W	24	6	DD	32°31'N 70°00'W	2	Kindley AFB, Bermuda
4. 510 NM radius of 25°30'N, 148°00'W	24	6	DD	25°30'N 148°00'W	2	Hickam AFB, Hawaii

EARTH ORBITAL AND DEEP SPACE

CONTINGENCY LANDING AREA

DESCRIPTION	ACCESS TIME HR A/C	A/C READINESS	A/C NO.	STAGING BASES
All area outside the launch site, launch abort, primary and secondary landing areas between 40°N 15°S. For earth orbital phase, latitude limits are 34°N and 34°S.	18	See Tab A to Appendix VII	2 2 2 2 2 1 2	Bermuda (May be released after TLI) Ascension Island Lajes/Moron (May be released after TLI) Mauritius Island Hickam AFB, Hawaii Andersen AFB, Guam (SAR Alert) Howard AFB, Canal Zone

DEEP SPACE TYPICAL SECONDARY LANDING AREA AND FORCE DEPLOYMENT

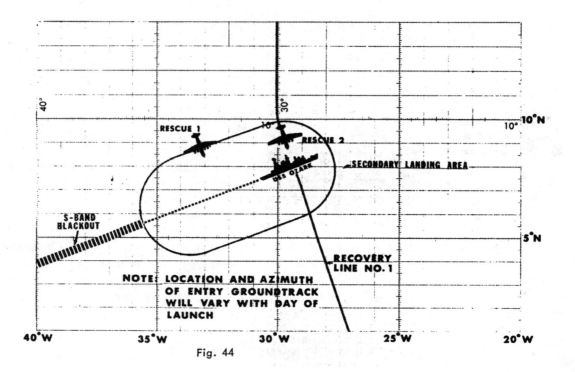

Fig. 44

TABLE 10

HC-130 MINIMUM ALERT POSTURE

STAGING BASE	LAUNCH TO PARKING ORBIT INSERTION	PARKING ORBIT INSERTION TO TLI	AFTER TLI*	REMARKS
Pease	Aircraft A airborne in Launch Abort Area	1 aircraft with 1/2 hr reaction time. Aircraft A or B can provide support while returning to home base	Aircraft can be released after TLI	
Kindley	Aircraft B airborne in Launch Abort Area		Aircraft can be released after TLI	
Lajes	Aircraft C airborne in Launch Abort Area	Aircraft returning to home base	Aircraft can be released after TLI	
Ascension	Not Required	1 aircraft with 2 hr reaction time	2 aircraft with 1/2 hr reaction time until TLI + 4 hrs	Aircraft can return to home base after TLI + 35 hours. Aircraft can be released at entry minus 37 hours if CM is still targeted to MPL.
Mauritius	Not Required	1 aircraft with 2 hr reaction time	1 aircraft with 6 hr reaction time until TLI plus 14 hrs	Aircraft can be released at entry minus 56 hours if CM still targeted to MPL.

*Reaction times are designed to provide required support during first few hours after TLI for any possible mission launched during the July launch window. After TLI the mission trajectory will have been established and more relaxed reaction times will be possible based on the minimum return time. These minimum return times will be passed to recovery forces as they are identified.

TABLE 9
RECOVERY SHIP LOCATIONS, DEEP SPACE PHASE

Launch Date	Mid-Pacific Line USS HORNET	Atlantic Ocean Line USS OZARK
16 July	03°00'S, 165°00'W	01°00'S 26°25'W
18 July	09°30'N, 171°10'W	11°00'N, 30°00'V
21 July	25°30'N, 175°00'W	24°00'N, 30°00'W

END-OF-MISSION PHASE

The normal end-of-mission (EOM) landing area will be selected on or near the MPL (line 4) located in the Mid-Pacific Ocean as shown in Figure 45. The latitude of the target point will depend on the declination of the moon at transearth injection and will be in the general range of 11°N to 29°N for the July launch window. The target point will normally be 1285 NM downrange of the entry point. Forces will be assigned to this area, as listed in Table 11, to meet the specified access and retrieval times. These forces will be on station not later than 10 minutes prior to predicted CM landing time.

If the entry range is increased to avoid bad weather, the area moves along with the target point and contains all the high probability landing points as long as the entry range does not exceed 2000 NM. Access and retrieval times quoted for the primary landing area will not apply if entry ranges greater than 2000 NM are flown during the mission.

The recovery forces in the primary landing area will be capable of meeting:

A maximum access time of 2 hours to any point in the area.

A maximum crew retrieval time of 16 hours to any point in the area.

TABLE 11

RECOVERY FORCE REQUIREMENTS

DEEP SPACE PHASE

MID-PACIFIC LINE 4

RECOVERY ZONES	RETRIEVAL TIME (HR) SHIP	ACCESS TIME (HR) A/C	SHIP		HC-130 AIRCRAFT	
			TYPE	POSITION	NO	STAGING BASES
A 125 NM circle centered 275 NM uprange A 125 NM circle centered 25 NM uprange from the target point connected by two tangential lines	24	14	CVS	At TP, latitude dependent on launch day	2	Hickam AFB, Hawaii. One A/C 200 NM uprange and 100 NM north of ground track. One A/C 200 NM downrange of TP and 100 NM north of ground track.

ATLANTIC OCEAN LINE 1

Same as Mid-Pacific Line	24	14	MCS-2	Same as Mid-Pacific Line	2	Ascension. One HC-130 200 NM uprange at TP and 100 NM north of ground track. One HC-130 abeam of TP and 50 NM north of ground track.

DEEP SPACE PRIMARY LANDING AREA

MPL DESCRIPTION	RETRIEVAL TIME (HR) SHIP	ACCESS TIME (HR) A/C	DEPLOYMENT					
			SHIP				AIRCRAFT	
			NO.	TYPE	POSITION	NO.	POSITION	
Same as Mid-Pacific line	Crew 16 CM 24	2	1	CVS	At TP, latitude dependent on launch day (as updated)	4 2 1	Helos (*) HC-130 (*) E-1B (AIR BOSS)	
						1	EC-135 (ARIA)	

TYPICAL PRIMARY LANDING AREA AND FORCE DEPLOYMENT

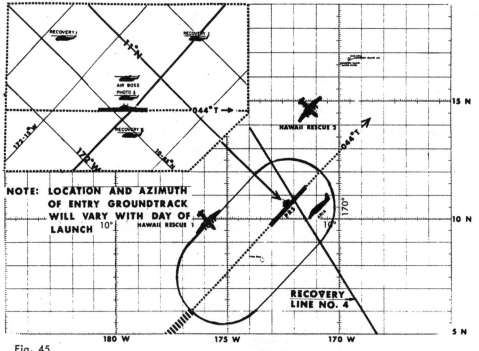

Fig. 45

A maximum CM retrieval time of 24 hours to any point in the area.

The recovery forces assigned to the primary landing area are:

USS HORNET will be on the EOM target point.

Three SARAH-equipped helicopters, each carrying a three-man swimmer team, to conduct electronic search are required. At least one of the swimmers on each team will be equipped with an underwater (Calypso) 35mm camera. NASA will furnish the equipment and film and will brief the swimmers concerning employment and coverage required.

One helicopter to carry photographers as designated by the NASA Recovery Team Leader assigned to USS HORNET in the vicinity of the target point.

One aircraft to function as communications relay, stationed overhead at the scene of action.

One fixed-wing or rotary-wing aircraft over USS HORNET to function as on-scene commander.

One HC-130 aircraft with operational AN/ARD-17 (Cook Tracker), 3-man pararescue team, and complete Apollo recovery equipment will be stationed 200 NM uprange from the target point and 100 NM north of the CM ground track at 25,000 feet.

One HC-130 aircraft with operational AN/ARD-17, 3-man pararescue team, and complete Apollo recovery equipment will be stationed 200 NM downrange from the target point and 100 NM north of the CM ground track at 25,000 feet.

Prior to CM reentry, one EC-135 Apollo Range Instrumentation Aircraft will be on station near the primary landing area for network support.

FLIGHT CREW

FLIGHT CREW ASSIGNMENTS

Prime Crew (Figure 46)

Commander (CDR) - Neil A. Armstrong (Civilian) Command Module Pilot (CMP) - Michael Collins (Lt. Colonel, USAF) Lunar Module Pilot (LMP) - Edwin E. Aldrin, Jr. (Colonel, USAF)

Backup Crew (Figure 47)

Commander (CDR) - James A. Lovell, Jr. (Captain, USN) Command Module Pilot (CMP) William A. Anders (Lt. Colonel, USAF) Lunar Module Pilot (LMP) - Fred Wallace Haise, Jr. (Civilian)

The backup crew follows closely the training schedule for the prime crew and functions in three significant categories. One, they are fully informed assistants who help the prime crew organize the mission and check out the hardware. Two, they receive nearly complete mission training which becomes a valuable foundation for later assignments as a prime crew. Three, should the prime crew become unavailable, they are prepared to fly as prime crew up until the last few weeks prior to launch. During the final weeks before launch, the flight hardware and software, ground hardware and software, and flight crew and ground crews work as an integrated team to perform ground simulations and other tests of the upcoming mission. It is necessary that the flight crew that will conduct the mission take part in these activities, which are not repeated for the benefit of the backup crew. To do so would add an additional costly and time consuming period to the prelaunch schedule, which for a lunar mission would require rescheduling for a later lunar launch window.

APOLLO 11 PRIME CREW

NEIL A. ARMSTRONG   MICHAEL COLLINS   EDWIN E. ALDRIN JR.

Fig. 46

APOLLO 11 BACK-UP CREW

JAMES A. LOVELL JR.   WILLIAM A. ANDERS   FRED W. HAISE JR.

Fig. 47

PRIME CREW BIOGRAPHICAL DATA

Commander (CDR)

NAME: Neil A. Armstrong (Mr.)

BIRTHPLACE AND DATE: Wapakoneta, Ohio; 5 August 1930.

PHYSICAL DESCRIPTION: Blond hair; blue eyes; height; 5 ft. 11 in.; weight: 165 lb.

EDUCATION: Received a Bachelor of Science degree in Aeronautical Engineering from Purdue University in 1955. Graduate School - University of Southern California.

ORGANIZATIONS: Associate Fellow of the Society of Experimental Test Pilots; Associate Fellow of the American Institute of Aeronautics and Astronautics; and member of the Soaring Society of America.

SPECIAL HONORS: Recipient of the 1962 Institute of Aerospace Sciences Octave Chanute Award; the 1966 AIAA Astronautics Award; the NASA Exceptional Service Medal; and the 1962 John J. Montgomery Award.

EXPERIENCE: Armstrong was a naval aviator from 1949 to 1952. In 1955 he joined NASA's Lewis Research Center (then NACA Lewis Flight Propulsion Laboratory) and later transferred to the NASA High Speed Flight Station (now Flight Research Center) at Edwards Air Force Base, California, as an aeronautical research pilot for NACA and NASA. In this capacity, he performed as an X-15 project pilot, flying that aircraft to over 200,000 feet and approximately 4000 miles per hour. Other flight test work included piloting the X-1 rocket airplane, the F-100, F-101, F-102, F104, F5D, B-47, the paraglider, the B-29 "drop" airplane, and others.

CURRENT ASSIGNMENT: Mr. Armstrong was selected as an astronaut by NASA in September 1962. He served as the backup Command Pilot for the Gemini 5 flight.

As Command Pilot for the Gemini 8 Mission, which was launched on 16 March 1966, he performed the first successful docking of two vehicles in space. The flight, originally scheduled to last 3 days, was terminated early due to a malfunctioning attitude system thruster, but the crew demonstrated exceptional piloting skill in overcoming this problem and bringing the spacecraft to a safe landing.
He subsequently served as backup Command Pilot for the Gemini 11 Mission and backup Commander for the Apollo 8 Mission. In his current assignment as Commander for the Apollo 11 Mission, he will probably be the first human to set foot on the moon.

Command Module Pilot (CMP)

NAME: Michael Collins (Lieutenant Colonel, USAF)

BIRTHPLACE AND DATE: Rome, Italy; 31 October 1930.

PHYSICAL DESCRIPTION: Brown hair; brown eyes; height: 5 ft. 11 in.; weight 165 lb.

EDUCATION: Received a Bachelor of Science degree from the United States Military Academy at West Point, New York, in 1952.

ORGANIZATIONS: Member of the Society of Experimental Test Pilots.

SPECIAL HONORS: Awarded the NASA Exceptional Service Medal, the Air Force Command Pilot Wings, and the Air Force Distinguished Flying Cross.

EXPERIENCE: Collins chose an Air Force career following graduation from West Point. He served as an experimental flight test officer at the Air Force Flight Test Center, Edwards Air Force Base, California. In that capacity, he tested performance, stability, and control characteristics of Air Force aircraft primarily jet fighters.

CURRENT ASSIGNMENT: Lt. Colonel Collins was one of the third group of astronauts named by NASA in October 1963. His first assignment was as backup Pilot for the Gemini 7 Mission.

As Pilot of the 3-day, 44-revolution Gemini 10 Mission, launched 18 July 1966, Collins shares with Command Pilot John Young in the accomplishments of that record-setting flight - a successful rendezvous and docking with a separately launched Agena target vehicle and, using the power of the Agena, maneuvering the Gemini spacecraft into another orbit for a rendezvous with a second, passive Agena. The spacecraft landed 2.6 miles from the USS GUADALCANAL and became the second in the Gemini Program to land within eye and camera range of a primary recovery ship.

He was assigned as Command Module Pilot on the prime crew for the Apollo 8 Mission but was replaced when spinal surgery forced a lengthy recuperation.

Lunar Module Pilot (LMP)

NAME: Edwin E. Aldrin, Jr. (Colonel, USAF)

BIRTHPLACE AND DATE: Montclair, New Jersey; 20 January 1930.

PHYSICAL DESCRIPTION: Blond hair; blue eyes; height: 5 ft. 10 in.; weight: 165 lb.

EDUCATION: Received a Bachelor of Science degree from the United States Military Academy at West Point, New York, in 1951 and a Doctor of Science degree in Astronautics from the Massachusetts Institute of Technology in 1963; recipient of an Honorary Doctorate of Science degree from Gustavus Adolphus College in 1967.

ORGANIZATIONS: Fellow of the American Institute of Aeronautics and Astronautics; member of the Society of Experimental Test Pilots, Sigma Gamma Tau (aeronautical engineering society), Tau Beta Pi (national engineering society), and Sigma Xi (national science research society); and a 32nd Degree Mason advanced through the Commandery and Shrine.

SPECIAL HONORS: Awarded the Distinguished Flying Cross with one Oak Leaf Cluster, the Air Medal with two Oak Leaf Clusters, the Air Force Commendation Medal, the NASA Exceptional Service Medal and Air Force Command Pilot Astronaut Wings, the NASA Group Achievement Award for Rendezvous Operations Planning Team, an Honorary Life Membership in the International Association of Machinists and Aerospace Workers, and an Honorary Membership in the Aerospace Medical Association.

EXPERIENCE: Aldrin was graduated third in a class of 475 from the United States Military Academy at West Point in 1951 and subsequently received his wings at Bryan, Texas in 1952.

He flew combat missions in F-86 aircraft while on duty in Korea with the 51st Fighter Interceptor Wing. At Nellis Air Force Base, Nevada, he served as an aerial gunnery instructor and then attended the Squadron Officers' School at the Air University, Maxwell Air Force Base, Alabama.

Following his assignment as Aide to the Dean of Faculty at the United States Air Force Academy, Aldrin flew F-100 aircraft as a flight commander with the 36th Tactical Fighter Wing at Bitburg, Germany. He attended MIT, receiving a doctorate after completing his thesis concerning guidance for manned orbital rendezvous, and was then assigned to the Gemini Target Office of the Air Force Space Division, Los Angeles, California. He was later transferred to the USAF Field Office at the Manned Spacecraft Center which was responsible for integrating DOD experiments into the NASA Gemini flights.

CURRENT ASSIGNMENT: Colonel Aldrin was one of the third group of astronauts named by NASA in October 1963. He has since served as backup Pilot for the Gemini 9 Mission, prime Pilot for the Gemini 12 Mission, and backup Command Module Pilot for the Apollo 8 Mission.

BACKUP CREW BIOGRAPHICAL DATA

Commander (CDR)

NAME: James A. Lovell, Jr. (Captain, USN)

BIRTHPLACE AND DATE: Cleveland, Ohio; 25 March 1928

PHYSICAL DESCRIPTION: Blond hair; blue eyes; height: 5 ft. 11 in.; weight: 170 lb.

EDUCATION: Attended the University of Wisconsin for 2 years; received a Bachelor of Science degree from the United States Naval Academy in 1952.

ORGANIZATIONS: Member of the Society of Experimental Test Pilots and the Explorers Club.

SPECIAL HONORS: Awarded the NASA Distinguished Service Medal, two NASA Exceptional Service Medals, the Navy Astronaut Wings, two Navy Distinguished Flying Crosses, and the 1957 FAI Delavauly and Gold Space Medals (Athens, Greece); co-recipient of the 1966 American Astronautical Society Flight Achievement Award and the Harmon International Aviation Trophy in 1966 and 1967; and recipient of the American Academy of Achievement Gold Plate Award and the New York State Medal for Valor in 1969.

EXPERIENCE: Lovell received flight training following graduation from Annapolis. He has had numerous assignments including a 4-year tour as a test pilot at the Naval Air Test Center, Patuxent River, Maryland. While there he served as program manager for the F4H weapon system evaluation. A graduate of the Aviation Safety School of the University of Southern California, he also served as a flight instructor and safety officer with Fighter Squadron 101 at the Naval Air Station, Oceana, Virginia.

CURRENT ASSIGNMENT: Captain Lovell was selected as an astronaut by NASA in September 1962. He has served as backup Pilot for the Gemini 4 flight and as backup Command Pilot for Gemini 9.

On 4 December 1965, he and Command Pilot Frank Borman were launched into space on the history-making Gemini 7 Mission. The flight lasted 330 hours 35 minutes, during which the following space "firsts" were accomplished: longest manned space flight; first rendezvous of two manned maneuverable spacecraft, as Gemini 7 was joined by Gemini 6; and longest multimanned space flight.

The Gemini 12 Mission, with Command Pilot Lovell and Pilot Edwin Aldrin, began on 11 November 1966. This 4-day 59-revolution flight brought the Gemini Program to a successful close.

Lovell served as Command Module Pilot for the epic 6-day journey of Apollo 8, man's maiden voyage to the moon - 21-27 December 1968. Apollo 8 was the first "manned spacecraft" to be lifted into near-earth orbit by a 7.5 million pound thrust Saturn V Launch Vehicle, and every aspect of the mission went smoothly from liftoff to landing.

Having completed three space flights, Captain Lovell holds the endurance record for time in space with a total of 572 hours 10 minutes.

Command Module Pilot (CMP)

NAME: William A. Anders (Lieutenant Colonel, USAF)

BIRTHPLACE AND DATE: Hong Kong; 17 October 1933.

PHYSICAL DESCRIPTION: Brown hair; blue eyes; height: 5 ft. 8 in.; weight: 145 lb.

EDUCATION: Received a Bachelor of Science degree from the United States Naval Academy in 1955 and a Master of Science degree in Nuclear Engineering from the Air Force Institute of Technology at Wright-Patterson Air Force Base, Ohio, in 1962.

ORGANIZATIONS: Member of the American Nuclear Society and Tau Beta Pi.

SPECIAL HONORS: Awarded the Air Force Commendation Medal, Air Force Astronaut Wings, the NASA

Distinguished Service Medal, and the New York State Medal for Valor.

EXPERIENCE: Anders was commissioned in the Air Force upon graduation from the Naval Academy. After Air Force flight training, he served as a fighter pilot in all-weather interceptor squadrons of the Air Defense Command.

After his graduate training, he served as a nuclear engineer and instructor pilot at the Air Force Weapons Laboratory, Kirtland Air Force Base, New Mexico, where he was responsible for technical management of radiation nuclear power reactor shielding and radiation effects programs.

CURRENT ASSIGNMENT: Lt. Colonel Anders was one of the third group of astronauts selected by NASA in October 1963. He has since served as backup Pilot for the Gemini 11 Mission.

Anders served as Lunar Module Pilot for the Apollo 8 Mission, which was launched 21 December 1968 and returned from its voyage around the moon on 27 December 1968. This epic 6-day flight was man's maiden voyage to the moon.

Lt. Colonel Anders has recently been nominated by the President to be Executive Secretary of the National Aeronautics and Space Council.

Lunar Module Pilot (LMP)

NAME: Fred Wallace Haise, Jr. (Mr.)

BIRTHPLACE AND DATE: Biloxi, Mississippi; 14 November 1933.

PHYSICAL DESCRIPTION: Brown hair; brown eyes; height: 5 ft. 9.5 in.; weight: 150 lb.

EDUCATION: Attended Perkinston Junior College (Association of Arts); received a Bachelor of Science with honors in Aeronautical Engineering from the University of Oklahoma.

ORGANIZATIONS: Member of the Society of Experimental Test Pilots, Tau Beta Pi, Sigma Gamma Tau, and Phi Theta Kappa.

SPECIAL HONORS: Recipient of the A. B. Hants Trophy as the outstanding graduate of class 64A from the Aerospace Research Pilot School in 1964; awarded the American Defense Ribbon and the Society of Experimental Test Pilots Ray E. Tenhoff Award for 1966.

EXPERIENCE: Mr. Haise began his military career in October 1952 as a Naval Aviation Cadet at the Naval Air Station in Pensacola, Florida.

He served as a tactics and all-weather flight instructor in the U.S. Navy Advanced Training Command at NAAS Kingsville, Texas, and was assigned as a U.S. Marine Corps fighter pilot to VMF-533 and 114 at MCAS Cherry Point, North Carolina, from March 1954 to September 1956. From March 1957 to September 1959, he was a fighter-interceptor pilot with the 185th Fighter Interceptor Squadron in the Oklahoma Air National Guard.

He served with the U.S. Air Force from October 1961 to August 1962 as a tactical fighter pilot and as Chief of the 164th Standardization-Evaluation Flight of the 164th Tactical Fighter Squadron at Mansfield, Ohio.

Haise was a research pilot at the NASA Flight Research Center at Edwards, California, before coming to Houston and the Manned Spacecraft Center; and from September 1959 to March 1963, he was a research pilot at the NASA Lewis Research Center in Cleveland, Ohio. During this time he authored the following papers which have been published: a NASA TND, entitled "An Evaluation of the Flying Qualities of Seven General -Aviation Aircraft;" NASA TND 3380, "Use of Aircraft for Zero Gravity Environment, May 1966;" SAE Business Aircraft Conference Paper, entitled "An Evaluation of General Aviation Aircraft Flying Qualities,"

30 March - 1 April 1966; and a paper delivered at the tenth symposium of the Society of Experimental Test Pilots, entitled "A Quantitative/Qualitative Handling Qualities Evaluation of Seven General-Aviation Aircraft," 1966.

CURRENT ASSIGNMENT: Mr. Haise is one of the 19 astronauts selected by NASA in April 1966. Haise served as backup Lunar Module Pilot for Apollo 8.

MISSION MANAGEMENT RESPONSIBILITY

Title	Name	Organization
Director, Apollo Program	Lt. Gen. Sam C. Phillips	NASA/OMSF
Director, Mission Operations	Maj. Gen. John D. Stevenson (Ret)	NASA/OMSF
Saturn V Vehicle Prog. Mgr.	Mr. Lee B. James	NASA/MSFC
Apollo Spacecraft Prog. Mgr.	Mr. George M. Low	NASA/MSC
Apollo Prog. Manager KSC	R. Adm. Roderick O. Middleton	NASA/KSC
Mission Director	Mr. George H. Hage	NASA/OMSF
Assistant Mission Director	Capt. Chester M. Lee (Ret)	NASA/OMSF
Assistant Mission Director	Col. Thomas H. McMullen	NASA/OMSF
Director of Launch Operations	Mr. Rocco Petrone	NASA/KSC
Director of Flight Operations	Mr. Christopher C. Kraft	NASA/MSC
Launch Operations Manager	Mr. Paul C. Donnelly	NASA/KSC
Flight Directors	Mr. Clifford E. Charlesworth	NASA/MSC
	Mr. Eugene F. Kranz	
	Mr. Glynn S. Lunney	
	Mr. Milton L. Windier	
Spacecraft Commander (Prime)	Mr. Neil A. Armstrong	NASA/MSC
Spacecraft Commander (Backup)	Captain James A. Lovell, Jr.	NASA/MSC

PROGRAM MANAGEMENT

NASA HEADQUARTERS

Office of Manned Space Flight
Manned Spacecraft Center
Marshall Space Flight Center
Kennedy Space Center

LAUNCH VEHICLE	SPACECRAFT	TRACKING AND DATA ACQUISITION
Marshall Space Flight Center The Boeing Co. (S-IC)	Manned Spacecraft Center	Kennedy Space Center
North American Rockwell Corp. (S-II)	North American Rockwell (LES, CSM, SLA)	Goddard Space Flight Center
McDonnell Douglas Corp. (S-IVB)	Grumman Aircraft Engineering Corp.	Department of Defense
IBM Corp. (IU)	(LM)	MSFN

ABBREVIATIONS AND ACRONYMS

AGS	Abort Guidance System
ALHT	Apollo Lunar Handtools
ALSCC	Apollo Lunar Surface Close-up Camera
AOL	Atlantic ocean Line
AOS	Acquisition of Signal
APS	Ascent Propulsion System (LM)
APS	Auxiliary Propulsion System (S-IVB)
ARIA	Apollo Range Instrumentation Aircraft
AS	Ascent Stage
AS	Apollo/Saturn
BIG	Biological Isolation Garment
BPC	Boost Protection Cover
CCATS	Communications, Command, and Telemetry System
CD	Countdown
CDH	Constant Delta Height
CDR	Commander
CES	Control Electronics System
CM	Command Module
CMP	Command Module Pilot
COI	Contingency Orbit Insertion
CRA	Crew Reception Area
CSI	Concentric Sequence Initiation
CSM	Command/Service Module
DOI	Descent Orbit Insertion
DPS	Descent Propulsion System
DS	Descent Stage
EASEP	Early Apollo Scientific ExperimentsPackage
ECS	Environmental Control System
EDS	Emergency Detection System
EDT	Eastern Daylight Time
EI	Entry Interface
EMU	Extravehicular Mobility Unit
EMS	Entry Monitor System
EOM	End-of-Mission
EPS	Electrical Power System
EPO	Earth Parking Orbit
EVA	Extravehicular Activity
EVCS	Extravehicular Communication System
GET	Ground Elapsed Time
GHe	Gaseous Helium
GNCS	Guidance, Navigation, and Control System
GOX	Gaseous Oxygen
H	Hybrid Trajectory
IMU	Inertial Measurement Unit
IS	Instrumentation System
IU	Instrument Unit
KSC	Kennedy Space Center
LC	Launch Complex
LCC	Launch Control Center
LCG	Liquid Cooling Garment
LES	Launch Escape System
LET	Launch Escape Tower
LH2	Liquid Hydrogen
LiOH	Lithium Hydroxide
LM	Lunar Module
LMP	Lunar Module Pilot
LOI	Lunar orbit Insertion
LOX	Liquid Oxygen
LPO	Lunar Parking Orbit
LRL	Lunar Receiving Laboratory
LRRR	Laser Ranging Retro-Reflector

LTA	Lunar Module Test Article
LV	Launch Vehicle
MCC	Midcourse Correction
MCC	Mission Control Center
MESA	Modularized Equipment Stowage Assembly
MOCR	Mission Operations Control Room
MOR	Mission Operation Report
MPL	Mid-Pacific Line
MQF	Mobile Quarantine Facility
MSC	Manned Spacecraft Center
MSFN	Manned Space Flight Network
MSS	Mobile Service Structure
NASCOM	NASA Communications Network
NM	Nautical Mile
OPS	Oxygen Purge System
PC	Plane Change
PDI	Powered Descent Initiation
PGNS	Primary Guidance and Navigation System
PLSS	Portable Life Support System
PRS	Primary Recovery Ship
PSE	Passive Seismic Experiment
PTP	Preferred Target Point
RCS	Reaction Control System
RR	Rendezvous Radar
R&D	Research and Development
RTCC	Real-Time Computer Complex
S&A	Safe and Arm
SAR	Search and Rescue
S/C	Spacecraft
SCS	Stabilization and Control System
SEA	Sun Elevation Angle
SEQ	Sequential System
SEQ	Scientific Equipment
SHe	Supercritical Helium
S-IC	First Stage
S-II	Second Stage
S-IVB	Third Stage
SLA	Spacecraft-LM Adapter
SLA	Secondary Landing Area
SM	Service Module
SPS	Service Propulsion System
SRC	Sample Return Container
SRS	Secondary Recovery Ship
SSR	Staff Support Room
SV	Space Vehicle
SXT	Sextant
SWC	Solar Wind Composition
TB	Time Base
TD&E	Transposition, Docking, and Ejection
T/C	Telecommunications
TEC	Transearth Coast
TEI	Transearth Injection
TLC	Translunar Coast
TLI	Translunar Injection
TPF	Terminal Phase Finalization
TPI	Terminal Phase initiation
T-time	Countdown time (referenced to liftoff time)
TV	Television
USB	Uniform S-band
VAB	Vehicle Assembly Building
VG	Velocity-to-be-Gained
VHF	Very High Frequency

NATIONAL AERONAUTICS AND SPACE ADMINISTRATION WASHINGTON, D.C. 20546

Post Launch
Mission Operation Report
No. M-932-69-11

24 July 1969

To: A/Administrator

From: MA/Apollo Program Director

Subject: Apollo 11 Mission (AS-506) Post Launch Mission Operation Report No. 1

The Apollo 11 Mission was successfully launched from the Kennedy Space Center on Wednesday, 16 July 1969 and was completed as planned, with recovery of the spacecraft and crew in the Pacific Ocean recovery area on Thursday, 24 July 1969. Initial review of the flight indicates that all mission objectives were attained. Further detailed analysis of all data is continuing and appropriate refined results of the mission will be reported in the Manned Space Flight Centers' technical reports.

Attached is the Mission Director's Summary Report for Apollo 11 which is hereby submitted as Post Launch Mission Operation Report No. 1. I recommend that the Apollo 11 Mission be adjudged as having achieved the primary objective of a manned lunar landing and return, and be considered a success.

Sam C. Phillips
Lt. General, USAF
Apollo Program Director

Approval:
George E. Mueller
Associate Administrator for Manned Space Flight

M-932-69-11 NASA OMSF PRIMARY MISSION OBJECTIVES FOR APOLLO 11

PRIMARY OBJECTIVE

Perform a manned lunar landing and return.

Sam C. Phillips	George E. Mueller
Lt. General, USAF	Associate Administrator
Apollo Program Director	for Manned Space Flight
Date: June 26, 1969	Date: June 26, 1969

RESULTS OF APOLLO 11 MISSION

Based upon a review of the assessed performance of Apollo 11, launched 16 July 1969 and completed 24 July 1969, this mission is adjudged a success in accordance with the objective stated above.

Sam C. Phillips	George E. Mueller
Lt. General, USAF	Associate Administrator
Apollo Program Director	for Manned Space Flight
Date: 24 July 1969	Date. July 24 1969

NATIONAL AERONAUTICS AND SPACE ADMINISTRATION WASHINGTON, D.C. 20546

MAO   24 July 1969

TO: Distribution

FROM: MA/Apollo Mission Director

SUBJECT: Mission Director's Summary Report, Apollo 11

INTRODUCTION

The Apollo 11 mission was planned to perform a manned lunar landing and return. Flight crew members were; Commander, Mr. N. A. Armstrong; Command Module Pilot, Lt. Col. M. Collins; Lunar Module Pilot, Col. E. E. Aldrin. Significant detailed mission information is contained in Tables 1 through 10. Initial review of the flight indicates that all mission objectives were attained (Reference Table 1). Table 2 lists achievements.

PRELAUNCH

The Apollo 11 countdown was accomplished with no unscheduled holds.

LAUNCH AND EARTH PARKING ORBIT

The Apollo 11 space vehicle was successfully launched from Kennedy Space Center, Florida, at 9:32 a.m. EDT on 16 July 1969. All launch vehicle stages performed satisfactorily, inserting the S-IVB/spacecraft combination into an earth parking orbit of 103 nautical miles (NM) circular — precisely as planned. (Refer to Table 4 for powered flight sequence of events). All systems operated satisfactorily.

Pre-TLI (translunar injection) checkout was conducted as planned and the second S-IVB burn was initiated on schedule. (Reference Table 6). All systems operated satisfactorily and all end conditions were nominal for the translunar coast on a free return trajectory.

TRANSLUNAR COAST

The Command/Service Module (CSM) was separated from the remainder of the orbital vehicle at about 3:17 GET, (hr:min ground elapsed time). The crew reported at 3:29 GET that CSM transposition and docking with the Lunar Module (LM)/Instrument Unit (IU)/S-IVB were complete. Ejection of the CSM/LM was successfully accomplished at about 4:17 GET and a SM Service Propulsion System (SPS) evasive maneuver was performed as planned at 4:40 GET. (Reference Table 6). All launch vehicle safing activities were performed as scheduled.

The S-IVB/IU slingshot maneuver was successful in avoiding spacecraft recontact, lunar impact and earth capture. The closest approach to the moon of 2340 nautical miles occurred at 78:50:34 (4:22:34 p.m. EDT, 19 July).

The accuracy with which the TLI maneuver was performed was such that midcourse correction number one (MCC-1) was not required.

An unscheduled 16-minute television transmission was recorded at the Goldstone station beginning at 10:32 GET. The tape was played back at Goldstone and transmitted to Houston beginning at 11:26 GET. An unscheduled 50-minute television transmission was accomplished at 30:28 GET, and a 36-minute scheduled TV transmission began at 33:59 GET.

MCC-2 was performed at 26:45 GET as planned and all SPS burn parameters were nominal (Reference Table 6). The accuracy of the MCC-2 calculation and performance was such that MCC-3 and MCC-4 were not necessary.

The crew initiated a 96-minute color television transmission at 55:08 GET. The picture resolution and general quality were exceptional. The coverage was outstanding, including the interior of both the CM and LM, and views of the exterior of the CM and the earth. Excellent views of the crew accomplishing probe and drogue removal, spacecraft tunnel hatch opening, LM housekeeping and equipment were obtained. The picture was transmitted live throughout North and South America, Japan, and Western Europe.

The spacecraft passed into the moon's sphere of influence at 61:39:55 GET. At that time, the distance from the spacecraft to earth was 186,436 NM and its distance from the moon was 33,823 NM. The velocity was 2,990 feet per second (fps) relative to earth and 3,775 fps relative to the moon.

LUNAR ORBIT

Lunar orbit insertion (LOI) was performed in two separate maneuvers using the SPS. The first maneuver, LOI-1, was initiated at 75:50 GET. The retrograde maneuver placed the spacecraft in a 168.8 x 61.3 NM elliptical orbit.

During the second lunar orbit, a scheduled live color television transmission was accomplished. Spectacular views of the lunar surface included the approach path to Lunar Landing Site 2.

After two revolutions and a navigation update, a second SPS retrograde burn (LOI-2) was made. The resulting orbit had an apolune of 65.7 NM and a perilune of 53.8 NM. LOI-1 and 2 burn parameters were nominal (Reference Table 7).

After LOI-2, the crew transferred to the LM and for about two hours performed various housekeeping functions, a voice and telemetry test, and an Oxygen Purge System check. LM functions and consumables quantities checked out very well. Additionally, both LM Hasselblad and Maurer cameras were checked and verified as being operational.

The Commander (CDR) and LM Pilot (LMP) re-entered the LM at approximately 95:20 GET to perform a thorough check of all LM systems in preparation for descent. Undocking of the LM from the CSM occurred at approximately 100:14 GET. Station-keeping was then initiated. At 100:40 GET, the SM Reaction Control System (RCS) was used to perform a small separation maneuver directed radially downward toward the center of the moon as planned (Reference Table 7).

DESCENT

The descent orbit insertion (DOI) maneuver was performed by a LM Descent Propulsion System (DPS) retrograde burn one-half revolution after LM/CSM separation, placing the LM in an elliptical orbit whose perilune was 8.5 NM (Reference Table 7).

The LM powered descent maneuver was initiated at perilune of the descent orbit. The time of powered descent initiate (PDI) was as planned. However, the position at which PDI occurred was about 4 NM downrange from that which was expected. This resulted in the landing point being shifted downrange about 4 NM.

During the final approach phase, the crew noted that the landing point toward which the spacecraft was headed was in the center of a large crater which appeared extremely rugged with boulders of 5 to 10 feet in diameter and larger. Consequently, the crew elected to fly to a landing point beyond this crater. This required manual attitude control and fine adjustments of the rate of descent plus high horizontal velocity to translate beyond the rough terrain area. As indicated above, the final landing point was estimated to be about four miles downrange from the center of the planned landing ellipse. Lunar landing occurred at 102:45:43 GET (4:17:43 P.M. EDT) (Reference Table 7 for powered descent parameters). Current estimate of landing point coordinates, based on analysis of all available data is 23.5° E and .64° N. This is approximately 20,800 feet west and 4,000 to 5,000 feet south of the center of the planned landing ellipse. Site altitude is estimated at approximately 8600 ft below the moon's mean radius.

LUNAR SURFACE ACTIVITIES

LM attitude on the surface was tilted 4.5° from the vertical, and yawed left about 13°. The crew indicated the landing site area contained numerous boulders of varying shapes and sizes up to 2 feet. The surface color varied from very light to dark gray. From his window view, the CDR reported seeing some boulders that were apparently fractured by engine exhaust. He reported that the surface of these boulders appeared to be coated light gray whereas the fractures were much darker. A hill was in sight at about ½ to 1 mile in front of the LM.

The crew indicated that they could immediately adapt to the 1/6 (earth) gravity in the LM and moved very easily in this environment. About two hours after landing, the crew requested that the extravehicular activity (EVA) be accomplished prior to the sleep period, or about 4½ hours earlier than originally scheduled. The rest period originally planned to occur prior to EVA was slipped until post-EVA and added to the second sleep period.

EXTRAVEHICULAR ACTIVITY

After the postlanding checks, the LM hatch was opened at 109:07:35 GET. As the CDR descended the LM ladder, he deployed the Modularized Equipment Stowage Assembly (MESA). The television camera mounted on the MESA access panel recorded his descent to the lunar surface. The CDR's first stop on the moon occurred at 109:24:15 GET (10:56:15 p.m. EDT). He made a brief check of the LM exterior, indicating that penetration of the footpads was only about 3 to 4 inches and collapse of the strut was minimal. He reported sinking approximately 1/8 inch into the fine, powdery surface material, which adhered readily to the lunar boots in a thin layer. There was no crater from the effects of the descent engine, and about one foot of clearance was observed between the engine bell and the lunar surface. He also reported that it was quite dark in the shadows of the LM which made it difficult for him to see his footing. During the EVA, a small microdot disk containing messages from numerous world leaders was left on the moon.

The CDR then collected a contingency sample of lunar soil from the vicinity of the LM ladder. He reported that although loose material created a soft surface, as he dug down six or eight inches, he encountered very hard cohesive material.

The CDR photographed the LMP's egress and descent to the lunar surface. The CDR and LMP then unveiled the plaque mounted on the strut behind the ladder and read its inscription to their worldwide television audience. Next, the CDR removed the TV camera from the Descent Stage MESA, obtained a panorama, and placed the camera on its tripod in position to view the subsequent surface EVA operations.

The LMP deployed the Solar Wind Composition experiment on the lunar surface in direct sunlight to the north of the LM as planned.

Subsequently, the crew erected a 3 x 5-foot American flag on an 8-foot aluminum staff. During the ensuing environmental evaluation, the LMP indicated that he had to be careful of his center of mass in maintaining balance. He noted that the LM shadow had no significant effect on his EMU temperature. The LMP also noted that his agility was better than expected. He was able to move about with great ease. Both crewmen indicated that their mobility throughout the EVA significantly exceeded all expectations. Also, indications were that metabolic rates were much lower than premission estimates.

A conversation between President Nixon and Armstrong and Aldrin was held. The conversation originated from the White House and contained congratulations and good wishes.

The CDR collected a Bulk Sample consisting of assorted surface material and selected rock chunks, and placed them in a Sample Return Container (SRC). Following the Bulk Sample collection, the crew inspected the LM and reported no discrepancies. The quads, struts, skirts, and antennas were satisfactory.

The Passive Seismic Experiment Package (PSEP) and Laser Ranging Retro Reflector were deployed south of the LM. Excellent PSEP data was obtained including detection of the crewmen walking on the surface and later, in the LM. The crew then collected more lunar samples until EVA termination including two core samples and about 20 pounds of discretely selected material. The LMP had to exert a considerable force to drive the core tubes an estimated six to eight inches deep.

Throughout the EVA, TV was useful in providing continuous observation for time correlation of crew activity with telemetered data and voice comments and in providing live documentation of this historically significant achievement. Lunar surface photography consisted of both still and sequence coverage using the Hasselblad camera, the Maurer data acquisition camera and the Apollo Lunar Surface Close-Up camera.

EVA termination film and sample transfer, LM ingress, and equipment jettison were accomplished according to plan. A rest period followed the post-EVA activities prior to preparation for liftoff. A record of lunar events is presented in Table 3.

ASCENT, RENDEZVOUS AND TRANSEARTH INJECTION

LM liftoff from the lunar surface occurred at 124:22 GET (1:34 p.m. EDT, 21 July) concluding a total lunar stay time of 21 hours 36 minutes. All lunar ascent and rendezvous maneuvers were nominal and terminated with CSM/LM docking at 128:03 GET. After transfer of the crew, samples and film to the CSM, the LM Ascent Stage was jettisoned at 130:10 GET. The LM ascent stage will remain in lunar orbit for an indefinite period of time. Subsequently, a small SM RCS separation maneuver placed the CSM in a 62.6 by 54.7 nautical mile orbit. (Reference Table 7 for maneuver parameters).

At 135:24 GET, the SPS injected the CSM into a transearth trajectory after a total time in lunar orbit of 59 hours 28 minutes (30 revolutions) (Reference Table 6). The TEI resulted in a transearth return time of about 60 hours.

TRANSEARTH COAST AND ENTRY

MCC-5 was initiated at 150:30 GET. The 10.8 second SM RCS burn produced a velocity change of 4.7 feet per second (Reference Table 6). An 18-minute television transmission was initiated at 155:36 GET and produced good quality pictures. The transmission featured crew demonstrations of the effect of weightlessness on food and water and brief scenes of the moon and earth. The accuracy of MCC-5 was such that MCC-6 and MCC-7 were not required.

The final color television broadcast was made at 177:32 GET. The 12½ minute transmission featured a sincere message of appreciation by each crew member to all people who helped make the Apollo 11 mission possible.

The crew awoke at 189:15 GET and initiated reentry preparations. CM/SM separation occurred at 194:49:19 GET and entry interface was reached at 195:03 GET.

Because of deteriorating weather in the nominal landing area, the aim point had been moved downrange 215 NM. Weather in the new landing area was excellent: visibility was 12 miles, wave height 3 feet, and wind 16 knots.

Visual contact of the spacecraft was reported at 195:06 GET. Drogue and main parachutes deployed normally. Landing occurred about 14 minutes after entry interface at 195:18:35 GET (12:50:35 EDT). The landing point was in the mid-Pacific, approximately 169:09°W longitude by 13:18° N latitude, about 13 NM from the prime recovery ship, USS HORNET. The CM landed in the Stable 2 position. Flotation bags were deployed to right the S/C into Stable 1 position at 195:25:10. The crew reported that they were in good condition.

ASTRONAUT RECOVERY OPERATIONS

Following landing, the recovery helicopter dropped swimmers who installed the flotation collar to the CM. A large, 7-man raft was deployed and attached to the flotation collar. Biological Isolation Garments (BIG's) were lowered into the raft, and one swimmer donned a BIG while the astronauts donned BIG's inside the CM. Two other swimmers moved upwind of the CM on a second large raft. The postlanding ventilation fan was turned off, the CM powered down, and the astronauts egressed and assisted the swimmer in closing the CM hatch. The swimmer then decontaminated all garments, the hatch area, the collar, and the area around the post landing vent valves.

The helicopter recovered the astronauts. After landing on the recovery carrier, the astronauts and a recovery physician entered the Mobile Quarantine Facility (MQF).

President Nixon, aboard USS HORNET, spoke to the crew members by intercommunications. He congratulated the Apollo 11 crew for this stupendous feat.

The flight crew, recovery physician and recovery technician will remain inside the MQF until it is delivered to the Lunar Receiving Laboratory (LRL) in Houston, Texas. This delivery is currently planned to occur on July 27.

CM RETRIEVAL OPERATIONS

After flight crew pickup by the helicopter, the CM was retrieved and placed in a dolly aboard the recovery ship. It was then moved to the MQF and mated to the transfer tunnel. From inside the MQF/CM containment envelope, the MQF engineer began post-retrieval procedures (removal of lunar samples, data, equipment, etc.), passing the removed items through the decontamination lock. The CM will remain sealed during RCS deactivation and delivery to the LRL.

The Sample Return Containers (SRC), film, data, etc. will be flown to Johnson Island by fixed wing aircraft from USS HORNET. The two SRC's will then be flown by separate aircraft to Houston for transport to the LRL.

MISSION SCIENTIFIC ACTIVITY

CONTINGENCY SAMPLE

A Contingency Sample was acquired early in the EVA in the immediate MESA area. The CDR indicated little difficulty in scooping material to a depth of six inches.

SOLAR WIND COMPOSITION EXPERIMENT

The Solar Wind Composition Experiment aluminum foil was deployed in the direct sunlight north of the LM. After an exposure for 77 minutes it was retracted and packaged in Sample Return Container number 2. The Principal Investigator has stated that exposure time of 75 minutes or higher would provide optimum results. There will be no data available from this experiment until after it is released from quarantine and returned to Switzerland for analysis.

BULK SAMPLE

A full sample bag was acquired in the Bulk Sample collection. The CDR collected a number of rock fragments of a wide variety of texture, color, size, angularity and consistency. The soil or "groundmass" was defined as being like very fine sand or silt; the best terrestrial analog according to the CDR is "powdered graphite".

CLOSEUP STEREO CAMERA

Closeup stereo photographs of various lunar surface areas were taken during LM inspection by the LMP.

HASSELBLAD SURFACE CAMERA

Extensive photographs were taken of the sampling activity and sample areas. Deployed experiments as well as the general topography were photographed.

EARLY APOLLO SCIENTIFIC EXPERIMENTS PACKAGE

Because of the LM landing attitude the Early Apollo Scientific Experiments Packages (EASEP) were manually removed from the LM; that is, the LM extendable booms were not used, The packages were deployed between the 9 and 10 o'clock positions referenced to the LM hatch (as 12 o'clock) at a distance of approximately 50 feet.

PASSIVE SEISMIC EXPERIMENT PACKAGE

Data acquisition began approximately four minutes after Passive Seismic Experiment PSEP) deployment or approximately 00:45 a.m. EDT, July 21. All engineering data followed preflight predictions during initial activity. The EVA and astronaut/LM activity were detected by the seismometer.

Telemetry received subsequent to LM lunar liftoff indicated that the Central Station thermal plate temperature experienced a substantial increase in rate of temperature rise. Preliminary analysis indicates a degradation of PSEP radiating surfaces of some 10 to 15 per cent.

The Principal Investigator has identified several lunar seismic events that are attributable to natural phenomena.

LASER RANGING RETRO-REFLECTOR

The Laser Ranging Retro-Reflector (LRRR) experiment is optimally designed for lunar night operation and has consequently not yet been acquired by any laser ranging stations.

DOCUMENTED SAMPLE

During the abbreviated Documented Sampling period two core tube samples of 6 to 8 inches in depth were obtained. Discrete samples were obtained north of the LM and around an elongated double crater to the south. Approximately 20 pounds of sample were obtained in this period and a number of photographs were taken of an 80 foot (diameter) by 15 foot (deep) crater containing rocks at the bottom.

The Contingency Sample, the two SRC's and film cassettes were successfully transferred to the LM and subsequently to the CM.

SYSTEMS PERFORMANCE

All launch vehicle systems performed satisfactorily throughout their expected lifetimes. There were no launch vehicle anomalies. All spacecraft systems continued to function satisfactorily throughout the mission. No major anomalies occurred. Those CSM and LM discrepancies which did occur are described in Tables 9 and 10, respectively. Temperatures and consumables usage rates remained generally within normal limits throughout the mission. Refer to Table 8 for a summary of end of mission consumables. Complete analyses of systems performance will be reported in subsequent Manned Space Flight Center engineering reports.

FLIGHT CREW PERFORMANCE

Flight crew performance was outstanding. All three crew members remained in excellent health. They made continued references to the high quality of the food.

The flight crew's (as well as ground crew's) good spirits and highly proficient performance were evident throughout the mission and were a major factor in the mission's enormously successful results. There are two quotes from Commander Armstrong that I believe reflect the spirit of this mission to the Apollo team:

"Houston, Tranquillity Base here — the Eagle has landed."

and

"One small step for man — one giant leap for mankind."

These statements truly reflect the significant way in which the lunar landing mission was executed as a team effort.

All information and data in this report are preliminary and subject to revision by the normal Manned Space Flight Center technical reports.

G.H. Hage

LIST OF TABLES

DETAILED OBJECTIVES AND EXPERIMENTS

24 July 1969   TABLE 1

DETAILED OBJECTIVES AND EXPERIMENTS

	ACCOMPLISHMENT SCHEDULE (DAY)								
	1	2	3	4	5	6	7	8	9
A. Contingency Sample collection.					X				
B. Lunar surface EVA operations.					X				
C. EMU lunar surface operations.					X				
D. Landing effects on LM.					X				
E. Lunar surface characteristics.					X				
F. Bulk Sample collection.					X				
G. Landed LM location.					X				
H. Lunar environment visibility.				X	X	X			
I. Assessment of contamination by lunar material.					X	X	X	X	X
S-031. Passive Seismic Experiment.					X				
S-078. Laser Ranging Retro-Reflector.					X				
S-080. Solar Wind Composition.					X				
S-059. Lunar Field Geology.					X				
L. Television coverage					X				
M. Photographic coverage.					X				

1. It is considered that accomplishment of the primary objective (manned lunar landing and return) qualified Apollo 11 as a success. The accomplishment of the Detailed Objectives and Experiments further enhanced the scientific and technological return of this mission.

2. Other major activities not listed as Detailed Objectives or Experiments:

Color TV LM Post Jettison Systems Lifetime Evaluation (emphasis on PGNCS operation without primary coolant loop). Passive Thermal Control Mode Test Visual Observation of Transient events at Aristarchus

3. All objectives were 100 per cent accomplished except as indicated below:

* Detailed Objective G: May be only partial, because it is not yet known if photographs of landed LM were taken by CM.

* Experiment S-078: Since ranging with the retro-reflector has not yet been accomplished, the degree of this experiment's success is not yet known.

* Experiment S-059: Carried as partial, because the environmental and gas analysis samples were not collected.

24 July 1969   TABLE 2

APOLLO 11 ACHIEVEMENTS

FIRST MANNED LUNAR LANDING AND RETURN.

FIRST LUNAR SURFACE EVA.

FIRST SEISMOMETER DEPLOYED ON MOON.

FIRST LASER REFLECTOR DEPLOYED ON MOON.

FIRST SOLAR WIND EXPERIMENT DEPLOYED ON MOON.

FIRST LUNAR SOIL SAMPLES RETURNED TO EARTH

SUCCESSFUL ACCOMPLISHMENT OF ALL MISSION OBJECTIVES.

SIXTH SUCCESSFUL SATURN V ON-TIME LAUNCH.

LARGEST PAYLOAD EVER PLACED IN LUNAR ORBIT.

FIRST LUNAR MODULE TEST IN TOTAL OPERATIONAL ENVIRONMENT.

ACQUISITION OF NUMEROUS VISUAL OBSERVATIONS, PHOTOGRAPHS AND TELEVISION OF SCIENTIFIC AND ENGINEERING SIGNIFICANCE.

FIRST OPERATIONAL USE OF MQF AND LRL.

<u>24 July 1969</u>   <u>TABLE 3</u>

APOLLO 11 RECORD OF LUNAR EVENTS

EVENT	Ground Elapsed Time Hr:Min:Sec	Greenwich Mean Time Hr:Min:Sec	Date
Lunar Orbit Insertion	75:55:52	17:27:52	19 July 69
Lunar Landing Time	102:45:43	20:17:43	20 July 69
LM Hatch Open	109:07:35	02:39:35	21 July 69
Armstrong Completely Out Of LM	109:19:16	02:51:16	21 July 69
Armstrong On Lunar Surface	109:24:15	02:56:15	21 July 69
Aldrin Out Of LM	109:39:00	03:11:00	21 July 69
Aldrin On Lunar Surface	109:43:15	03:15:15	21 July 69
Aldrin Inside LM	111:29:39	05:01:39	21 July 69
Armstrong Inside LM (Hatch Closed)	111:39:12	05:11:12	21 July 69
LM Liftoff From Lunar Surface	124:22:00	17:54:00	21 July 69
Transearth Injection	135:23:42	04:55:42	22 July 69

EVENT	Duration Hr:Min:Sec
Stay On Lunar Surface	21:36:17
Stay Outside LM	
Armstrong	02:31:37
Aldrin	01:50:24
Lunar Orbit	59:27:50
Weight Landed On Moon	15,897 Pounds
Weight Lifted From Moon	10,821 Pounds

<u>24 July 1969</u>   <u>TABLE 4</u>

APOLLO 11 POWERED FLIGHT SEQUENCE OF EVENTS

EVENT	*PLANNED (GET) HR:MIN:SEC	ACTUAL (GET) HR:MIN:SEC
Range Zero (09:32:00.0 EDT)	00:00:00.0	00:00:00.0
Liftoff Signal (TB-1)	00:00:00.6	00:00:00.6
Pitch and Roll Start	00:00:13.8	00:00:12.4
Roll Complete	00:00:31.8	00:00:31.1
S-IC Center Engine Cutoff (TB-2)	00:02:15.3	00:02:15.2
Begin Tilt Arrest	00:02:40.8	00:02:40.0
S-IC Outboard Engine Cutoff (TB-3)	00:02:41.1	00:02:41.6
S-IC/S-II Separation	00:02:41.8	00:02:42.3
S-II Ignition (Engine Start Command)	00:02:42.5	00:02:43.0
S-II Second Plane Separation	00:03:11.8	00:03:12.3
Launch Escape Tower Jettison	00:03:17.5	00:03:17.9
S-II Center Engine Cutoff	00:07:40.1	00:07:40.6
S-II Outboard Engine Cutoff (TB-4)	00:09:11.7	00:09:08.2
S-II/S-IVB Separation	00:09:12.5	00:09:09.0
S-IVB Ignition (Engine Start Command)	00:09:12.7	00:09:09.2
S-IVB Cutoff (TB-5)	00:11:39.5	00:11:39.3
Earth Parking Orbit Insertion	00:11:49.5	00:11:49.3
Begin S-IVB Restart Preparations (TB-6)	02:34:37.3	02:34:38.2
Second S-IVB Ignition	02:44:15.3	02:44:16.2
Second S-IVB Cutoff (TB-7)	02:50:04.1	02:50:03.0
Translunar Injection	02:50:14.1	02:50:13.0

*Prelaunch planned times are based on MSFC launch Vehicle operational trajectory.

24 July 1969   TABLE 5

APOLLO 11 MISSION SEQUENCE OF EVENTS

EVENT	*PLANNED (GET) HR:MIN:SEC	ACTUAL (GET) HR:MIN:SEC
Range Zero (09:32:00 EDT)	00:00:00	00.00:00
Earth Parking Orbit Insertion	00:11:49	00:11:49
Second S-IVB Ignition	02:44:15	02:44:16
Translunar Injection	02:50:14	02:50:13
CSM/S-IVB Separation, SLA Panel Jettison	03:15:00	03:17:-
CSM/LM Docking Complete	03:25:00	03:29:-
Spacecraft Ejection from S-IVB	04:09:45	04:17:13
Spacecraft Evasive Maneuver	04:39:45	04:40:01
S-IVB Slingshot Maneuver	05:02:03	Not available
Midcourse Correction -1	11:45:00	Not Performed
Midcourse Correction -2	26:45:00	26:44:58
Midcourse Correction -3	53:55:00	Not Performed
Midcourse Correction -4	70:55:00	Not Performed
LOI-1 (Lunar Orbit Insertion) Ignition	75:54:28	75:49:50
LOI-2 Ignition	80:09:30	80:11:36
LM Undocking from CSM	100:13:38	100:13:38
CSM Separation Maneuver	100:39:50	100:39:50
LM Descent Orbit Insertion	101:38:48	101:36:14
Powered Descent Initiation	102:35:13	102:33:04
Lunar Landing	102:47:11	102:45:43
Plane Change Maneuver (CSM)	107:05:33	Not Performed
Crew Egress for Lunar Surface Operations	112:40:00	109:07:35
Crew Ingress	115:10:00	111:39:12
LM Liftoff	124:23:26	124:22:00
Coelliptic Sequence Initiate Maneuver	125:21:19	125:19:35
Plane Change Maneuver (LM)	125:50:28	Not Performed
Constant Differential Height Maneuver	126:19:37	126:17:46.
Terminal Phase Initiate Maneuver	126:58:08	127:03:31
Terminal Phase Finalize Maneuver	127:40:38	127:45:54
CSM/LM Docking	128:00-00	128:03:-
LM Jettison	131:53:05	130:09:55
CSM Separation Maneuver	131:53:05	130:30:00
Transearth Injection (Ignition)	135:24:34	135:23:42
Midcourse Correction -5	150:24:00	150:29:55
Midcourse Correction -6	172:00:00	Not performed
Midcourse Correction -7	192:06:00	Not performed
CM/SM Separation	194:50:00	194:49:19
Entry Interface (400,000 feet)	195:05:04	195:03:06
Landing	195:19:06	195:18:35

*Prelaunch planned times are based on MSFC Launch Vehicle Operational Trajectory and MSC Spacecraft Operational Trajectory

APOLLO 11 TRANSLUNAR AND TRANSEARTH MANEUVER SUMMARY

TABLE 6

Date: 24 July 1969

MANEUVER	GROUND ELAPSED TIME (GET) AT IGNITION (hr:min:sec) PRE-LAUNCH PLAN	REAL-TIME PLAN	ACTUAL	BURN TIME (seconds) PRE-LAUNCH PLAN	REAL-TIME PLAN	ACTUAL	VELOCITY CHANGE (feet per second - fps) PRE-LAUNCH PLAN	REAL-TIME PLAN	ACTUAL	GET OF CLOSEST APPROACH / HT. (NM) CLOSEST APPROACH PRE-LAUNCH PLAN	REAL-TIME PLAN	ACTUAL
TLI (S-IVB)	2:44:15.3	2:44:16.2	2:44:16.2	349.5	347.5	347.3	10451.2	10435.9	10441.0	75:24:06.1 / 346.5	75:04:28 / 854.1	75:16:24 / 701.9
Evasive Maneuver (SPS)	4:39:44.9	4:39:44.9	4:40:01.0	2.8	2.8	3.4	19.7	19.7	19.7	75:57:39.4 / 59.8	75:40:35.2 / 335.0	75:38:22 / 179.7
MCC-1 (SPS)	11:45:00	11:30:00	N.P. (Not performed)	0.0	2.4	N.P.	0.0	17.3	N.P.	75:57:39.4 / 59.8	75:53:49.0 / 60.0	75:53:46 / N.P.
MCC-2 (SPS)	26:45:00	26:44:58	26:44:58	2.8	3.0	2.9	0.0	21.3	20.9	75:57:34.4 / 59.8	75:53:49 / 60.0	75:53:46 / 62.8
MCC-3	53:55:00	53:55:00	N.P.	0.0	8.0	N.P.	0.0	0.0	N.P.	75:57:34.4 / 59.8	75:53:49 / 60.0	N.P.
MCC-4	70:55:00	70:55:00	N.P.	0.0	21.6	N.P.	0.0	2.6	N.P.	75:51:34.4 / 59.8	75:53:49 / 60.0	N.P.
LUNAR ORBIT MANEUVERS	Lunar orbit maneuvers are summarized on a separate table.											
										GET ENTRY INTERFACE (EI) / VELOCITY (fps) AT EI / FLIGHT PATH ANGLE AT EI		
TEI (SPS)	135:24:33.8	135:23:41.6	135:23:42.0	149.1	147.9	150.0	3292.7	3283.6	3278.8	195:05:03.5 / 36194.3 / -6.50	195:05:03.5 / 36194.3 / -6.50	195:03:08 / 36194.3 / -6.46 No Entry Vac. Per. over 66 NM
MCC-5	150:24:00	150:29:54.5	150:29:54.5	0.0	11.0	10.8	0.0	4.8	4.7	192:05:03.5 / 36194.3 / -6.50	195:03:06 / 36194.3 / -6.51	195:03:04 / 36194.3 / -6.51
MCC-6	172:00:00	172:00:00	N.P.	0.0	1.1	N.P.	0.0	0.4	N.P.	195:05:03.5 / 36194.3 / -6.50	195:03:04 / 36194.3 / -6.51	N.P.
MCC-7	192:06:00	192:06:00	N.P.	0.0	.5	N.P.	0.0	.1	N.P.	195:05:03.5 / 36194.3 / -6.50	195:03:05 / 36194.3 / -6.50	N.P.

APOLLO 11 LUNAR ORBIT MANEUVER SUMMARY

TABLE 7

Date: 24 July 1969

MANEUVER	GROUND ELAPSED TIME (GET) AT IGNITION (hr:min:sec)			BURN TIME (seconds)			VELOCITY CHANGE (feet per second - fps)			APOLUNE/PERILUNE RESULTANT (NAUTICAL MILES)		
	PRE-LAUNCH PLAN	REAL-TIME PLAN	ACTUAL	PRE-LAUNCH PLAN	REAL-TIME PLAN	ACTUAL	PRE-LAUNCH PLAN	REAL-TIME PLAN	ACTUAL	PRE-LAUNCH PLAN	REAL-TIME PLAN	ACTUAL
Lunar Orbit Insertion	75:54:28.4	75:49:49.6	75:49:49.6	358.9	362.1	362.1	2924.1	2917.3	2917.5	169.8/59.7	169.1/61.1	168.8/61.3
Lunar Orb Circularization	80:09:29.7	80:11:36.0	80:11:36.0	16.4	17.0	17.0	157.8	159.2	158.8	65.7/53.6	65.7/53.7	65.7/53.8
CSM/LM Separation	100:39:50.4	100:39:50.0	100:39:50	8.0	8.0	8.2	2.5	2.5	2.6	63.1/55.6	64.0/56.0	63.7/55.8
Descent Orbit Insertion	101:38:48.0	101:36:14.1	101:36:14.1	28.0	29.8	29.8	74.0	76.4	76.4	60.0/8.2	57.2/8.5	57.2/8.5
Powered Descent Initiation	102:35:13.0	102:33:04.4	102:33:04.4	714.0	712.7	712.6	6775.0	6776.0	6775.8	0.0/0.0	0.0/0.0	0.0/0.0
(CSM) Plane Change	107:05:33.4	106:05:00	N.P.	.8	.8	N.P.	16.6	15.0	N.P.	63.1/55.6	64.0/56.0	N.P.
Ascent	124:23:26.0	124:22:00.0	124:22:00.0	437.9	439.4	439.9	6060.2	6070.2	6070.1	45.0/9.9	45.2/9.0	45.2/9.0
Coelliptic Sequence Initiate	125:21:19.1	125:19:34.7	125:19:34.7	44.8	48.5	47.0	49.4	53.2	51.5	45.7/44.9	47.1/45.5	48.6/45.3
(LM) Plane Change	125:50:28.0	126:12:33	N.P.	0.0	1.0	N.P.	0.0	.2	N.P.	45.7/44.9	47.0/45.5	N.P.
Constant Delta Altitude	126:19:37.0	126:17:46.0	126:17:46.0	2.0	18.2	18.1	4.5	20.0	19.9	45.1/42.8	47.0/40.9	47.0/40.9
Terminal Phase Initiate	126:58:08.4	126:57:00	127:03:30.8	22.2	22.7	22.8	24.6	25.1	25.3	61.2/42.5	61.1/43.9	61.2/43.9
Terminal Phase Finalize	127:40:37.7	127:39:34.2	127:45:54	28.3	28.4	28.4	31.4	31.5	31.4	59.5/59.0	62.6/56.6	62.2/56.6
CSM/LM Separation	131:53:04.7	130:30:00	130:30:00	3.2	6.5	7.1	1.0	2.0	2.2	59.6/59.0	62.6/54.7	62.6/54.7

24 July 1969   TABLE 8

APOLLO 11 CONSUMABLES SUMMARY AT END OF MISSION

CONSUMABLE		LAUNCH LOAD	PRELAUNCH PLANNED REMAINING	ACTUAL REMAINING
CM RCS PROP (POUNDS/PERCENT)	U	208/100	169/81	Not Available
SM RCS PROP (POUNDS/PERCENT)	U	1,225/100	629/51	622/51
SPS PROP (POUNDS/PERCENT)	TK	40,600/100	4,304/11	4467/11
SM HYDROGEN (POUNDS/PERCENT)	U	51.6/100	14.6/28	16.8/60
SM OXYGEN (POUNDS/PERCENT)	U	601.5/100	204/34	238/76
LM RCS PROP (POUNDS/PERCENT)	U	5491/100	*296/54	*200/37
LM DPS PROP (POUNDS/PERCENT)	U	17,921/100	** 911/5.1	** 444/2.5
LM APS PROP (POUNDS/PERCENT)	U	5,177/100	*211/4.1	* 411/3.0
LM A/S OXYGEN (POUNDS/PERCENT)	T	4.74/100	*2.29/48	* 3.8/80
LM D/S OXYGEN (POUNDS/PERCENT)	T	49.4/100	**25.0/51	** 30.9/63
LM A/S WATER (POUNDS/PERCENT)	T	84.4/100	*35.9/42	*46.0/54
LM D/S WATER (POUNDS/PERCENT)	T	214.4/100	**39.6/19	**70.0/33
LM A/S BATTERIES (AMP-HRS/PERCENT)	T	592/100	*162/27	*337/57
LM D/S BATTERIES (AMP-HRS/PERCENT)	T	1,600/100	*462/29	**541/34

U - Usable Quantity   *   At LM jettison
TK - Tank Quantity   ** At LM liftoff from moon
T - Total Quantity

24 July 1969   TABLE 9

COMMAND/SERVICE MODULE 107 DISCREPANCY SUMMARY

* PRIMARY AND SECONDARY ISOLATION VALVES ON QUAD B INDICATED BARBERPOLE AT PYRO FIRE FOR SEPARATION.

* CONDENSER EXIT TEMPERATURE ON FUEL CELL 2 SUDDENLY DROPS ABOUT 1° TO 2°F EVERY 5 MINUTES.

* ECS O2 FLOW TRANSDUCER READS .2 - .3 LBS/HR DURING THE CABIN O2 ENRICHMENT PURGE. THE READING SHOULD BE PEGGED HIGH (APPROXIMATELY .981 LBS/HR) DURING THIS CONDITION.

* UNEXPLAINED O2 FLOW MASTER ALARM OCCURRED AT 55:00 GET, WITH THE CLOSING OF THE DIRECT O2 VALVE, AND FOLLOWED AN EXPECTED ALARM.

* SPS GN2 PRESSURE INDICATED APPROXIMATELY 300 PSI LOWER THAN EXPECTED DURING LOI 1 BURN.

* PRIMARY GLYCOL MIXING VALVE DID NOT CONTROL MIXED TEMPERATURE TO 45° $\pm$ 3°F

* ONE HEATER IN CRYOGENIC OXYGEN TANK 2 FAILED BEFORE LAUNCH.

* CMP IMPEDANCE PNEUMOGRAPH (RESPIRATION) SIGNAL LOST.

24 July 1969   TABLE 10

LUNAR MODULE 5 DISCREPANCY SUMMARY

MISSION TIMER STOPPED AT 907:34:47 AND WHEN CREW RESET TIMER IT READ 902:34:47. CIRCUIT BREAKER WAS RECYCLED AND TIMER WENT TO ALL NINES. WOULD NOT START COUNTING.

POST PDI BURN DPS FUEL INTERFACE PRESSURE UNDERWENT RAPID PRESSURE RISE TO OFF SCALE READING. FUEL LINE FROZE DURING SHe VENTING TRAPPING FUEL BETWEEN PRE-VALVE AND SHe HEAT EXCHANGER CAUSING PRESSURE RISE.

DURING DESCENT, THE STEERABLE ANTENNA WAS NOT ABLE TO LOCK-UP PROPERLY ON THE MSFN. REACQUISITION WAS TRIED SEVERAL TIMES PRIOR TO PDI WITHOUT SUCCESS. AFTER THE YAW MANEUVER, PROPER ACQUISITION WAS OBTAINED.

THE CREW REPORTED THAT THE KNOB ON THE ASCENT ENGINE ARM CIRCUIT BREAKER WAS BROKEN.

AT APPROXIMATELY 126 HOURS THE CO2 PARTIAL PRESSURE READ HIGH CAUSING THE CREW TO SELECT THE SECONDARY Li OH CANISTER. THE READINGS REMAINED ERRATIC. THE PRIMARY CANISTER WAS AGAIN SELECTED. THE READINGS REMAINED ERRATIC AND CAUSED THE CAUTION AND WARNING TO BE ACTIVATED.

SUBSEQUENT TO ERRATIC CO2 SENSOR READINGS AT APPROXIMATELY 126 HOURS THE CREW REPORTED WATER IN ONE SUIT.

RENDEZVOUS RADAR TEMPERATURE STABILIZED 7° BELOW ACCEPTANCE TEST HEATER ACTIVATION LEVEL. RADAR OPERATION WAS NOT AFFECTED.

APOLLO NEWS CENTER HOUSTON, TEXAS
APOLLO 11 Postflight Crew Press Conference
August 12, 1969 10:00 a.m.

Neil A. Armstrong Commander
Edwin E. Aldrin. Jr. Lunar Module Pilot
Michael Collins   Command Module Pilot
Julian Scheer PAO

SCHEER — Ladies and gentlemen welcome to the Manned Spacecraft Center. This is an Apollo 11 press conference. The format today will consist of a 45 minute presentation by the Apollo 11 crew, followed by questions and answers. At this time I would like to introduce the Apollo 11 crew, astronauts Neil Armstrong, Michael Collins, Edwin Aldrin. Neil.

ARMSTRONG — It was our pleasure to participate in one great adventure. It's an adventure that took place not just in the month of July, but rather one that took place in the last decade. We all here and the people listening in today had the opportunity to share that adventure over its developing and unfolding in the past months and years. It is our privilege today to share with you some of the details of that final month of July that was certainly the highlight for the three of us of that decade. We're going to divert a little bit from the format of past press conferences and talk about the things that interested us most and particularly the things that occurred on and about the moon. We will use a number of films and slides which most of you have already seen and with the intent of pointing out some of the things that we observed on the spot which may not be obvious to those of you who are looking at them here from the surface of earth. The flight as you know started promptly and I think that was characteristic of all the events of the flight. The Saturn gave us one magnificent ride. Both into earth orbit and on a trajectory to the moon.(Photo. 1) Our memory of that actually differs a little from the reports that you have heard from the previous Saturn V flights and the previous flight served us well in preparation for this flight in the boost as well as the

subsequent phases. We would like to skip directly to the translunar coast phase and remind ourselves of the chain of events - that long chain of events that actually permitted the landing. Starting with the undockings - the transposition and docking sequence.

COLLINS — This was our first look at the magnificent machinery which had been behind us until this point. The booster - of course the first and second stages had long since separated, but this shows the LM - that's the LM side - the third stage the S-IVB, after the translunar injection burn. (Photo 2.) This maneuver was an interesting combination of manual and automated techniques in that we programmed onboard computer turn around and then these final maneuvers were made completely manually As I approached the LM I had an easy time because I had a docking target (Photo 3.)which is clearly visible here, which allowed me to align the probe and the drogue which is the dark spot which you see on the upper right. During this time I also checked out the proper vehicle response to my strict inputs and here shortly you'll see the actual docking somewhat speeded up. As a point of contact and in just a second you'll see a second - right there - a second indication of the retrack cycle which the 12 docking latches made.

Photo 3

ALDRIN — We made 2 entries into the lunar module. This is the first view of the inside of this. (Photo 4.) The final activation was made on the day of power descent and the 2 previous days when we entered removed the probe and drogue and found that we had a rather long tunnel between the two vehicles and entering the lunar module one has to do a slight flip maneuver or a half gainer to get into position, for the lunar module is in a sense up-side down relative to the command module.

COLLINS — This is in lunar orbit showing the separation of the lunar module from the command module as viewed through my window. (Photo 5.) This was a busy time for me in that I was taking these motion pictures through the right hand window at the same time as I was taking still photos through the left hand window and also flying my vehicle - and probably poorly - and taking a close look at the LM as he turned around. My most important job here was to make sure that all his landing gear was down and properly locked prior to his descent and touch down. This is his yaw maneuver and the white dots that you see are the landing gear pads. This gives you a better idea the details available with the 70 MM. Of course this is a still and shows the LM either right side up or upside down. I'm not sure which. It looks more

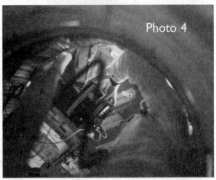

Photo 4

to me a praying mantis than it does a first class flying machine in this view, but it was a beautiful piece of machinery. The landing gear at the top and you can see the probes which indicate lunar contact is these thin wires extending upward from the landing gear.

Photo 5

ALDRIN — Of course before we could undock as is shown in this picture we had to complete the activation. Now the day before we undocked we entered the LM and went through an entire switch configuration check and we exercised the various communication modes. In retrospect since we did have a little bit of communication problems on the following day during power descent we would recommend that we might take a more thorough check of this on the day before descent. On the day that we did finally enter the LM for the landing maneuver we went through a staggered sequence of suiting and we found that with all the simulations that we had run back here in Houston - or with Houston tied with our simulations in the Cape that we were quite confident that we would be able to complete this LM activation in the given time period which was approximately 4 hours. We managed to get 30 minutes ahead of the time and it allowed us to get a more accurate platform alignment check at one point. After the undocking maneuver we went through a brief radar check and then the command module executed a 2 foot per second maneuver away from us so that we would both be able to independently exercise our guidance systems through a star alignment

Photo 6

check which we did following this separation maneuver. (Photo 6.) Soon after we were in the vicinity close to the landing sight and you can see at this point the command module is traveling right over the center of our targeted point. It's approaching what we call the Cats Paw. Following this separation maneuver on the back side of the moon we made a descent orbit insertion which was slightly over 70 feet per second that lowers our altitude down to 50 thousand feet. We had 2 guidance systems working for us. They behaved perfectly. Both of them agreed extremely closely as to the results of this maneuver. Following this we used the radar to confirm the actual departure rate from the command module.

Photo 7

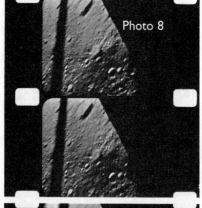

Photo 8

Photo 9

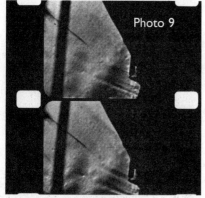

Photo 10

ARMSTRONG — This is a view of the descent trajectory area as viewed through the LM window during our activation. (Photo 7.) In the bottom right of the photograph is the crater Maskelyne and the bottom center is the mountain called Boot Hill. Immediately above Boot Hill is a small sharp-rimmed crater called Maskelyne W which was the crater that we used to determine our downrange and cross range position here prior to completing the final phases of the descent. The landing area itself is in the smooth area at the top of the picture just before we arrive at the shadow or what is called terminator. We had seen a number of pictures from Apollo's 8 and 10 which gave us an excellent understanding of the ground track over which we would pass during the descent. We are now looking out the right-hand window of the crater and there is Maskelyne W. It was approximately 2 to 3 seconds late and gave us the clue that we would probably land somewhat long. After completing those position checks we rolled over face up so that the landing radar could lock on the ground and confirm our actual altitude. Now, this picture doesn't show it, but at this phase in the trajectory we were looking out directly at the planet earth. In the final phases of descent after a number of program alarms, we looked at the landing area and found a very large crater just at the very left top corner of the picture. (Photo 8.) The camera is located in the right window and looks to the right and it just barely sees this boulder field that we are passing over right now. We are at 400 feet and those boulders are about 10 feet across. This is the area which we decided we would not go into; we extended the range downrange and saw this crater which we passed over, this 80-foot crater, in the final phases of descent and later took some pictures of. Now, you can see the exhaust dust being kicked up by the engine (Photo 9.) and this was some concern, in that it degraded our ability to determine not only our altitude and altitude-grade in the-final phases, but also, and probably more importantly, our translational velocities over the ground. It's quite important not to stub your toe during the final phases of touchdown. Once settled on the surface, the dust settled immediately and we had an excellent view of the area surrounding the LM. We saw a crater surface, pock-marked with craters up to 15, 20, 30 feet and many smaller craters down to diameter of 1 foot and, of course, the surface was very fine-grained. (Photo 10.) We could tell that from our view out the window, but there were a surprising number of rocks of all sizes.

ALDRIN — This is the view out the right window. (Photo 11.) Up close to the horizon you see a boulder field that was probably deposited by some of the impacts in the craters that were behind us. You see, most of the craters have rounded edges, however, there is a variation in the age of these as we can tell by the sharpness of the edge of the crater. The immediate foreground area we will see more pictures of later. It was relatively flat terrain in contrast to some of the more rolling terrain that we could see out the front window and out the left window. This is the view looking forward along where the shadow of the LM is cast on the surface (Photo 12.) - and we see a zero-phased glow around the upper portion of the LM. The general color of the terrain looking down sun was a very light tannish color. This blended as we looked more cross-sun to sharper, well-defined features and more of a gray color. During the initial time period after touchdown, we went to various sequences to prepare us for immediate abort or liftoff, if we found that this was necessary. We found that we had to vent the fuel and oxidizer manifolds a good bit earlier than we had thought. We went through these various checks and prepared for one liftoff that would occur about 21 minutes after the beginning of powered descent. The ground gave us a stay during this period

Photo 11

and we did not have to make use of that. We then proceeded at that point into our simulated countdown which consisted of checking our guidance systems. We made use of a gravity align feature where the inertial platform of the primary guidance would defuse the gravity vector to determine the local vertical. We then compared this with the alignments that we had previously. We also made use of the stars through the telescope in aligning a cross hair by rotating the field of view so the cross hair superimposed on the star - this would give us the angular measurement of the star within the field of view of the telescope. We then determined the distance out by aligning another radical spiral on this. We went through an averaging technique on board and then fed this information into the computer and this came up with our various alignment checks. This was all in preparation for a possible liftoff that would occur about two hours after touchdown as Mike and Columbia came over for the first revolution. The ground network gave us a stay and we continued through the remainder of the checklist in our simulator countdown and at this point we terminated and powered down the systems aboard the spacecraft and went into an eat period.

Photo 12

ARMSTRONG — A number of experts had, prior to the flight, predicted that a good bit of difficulty might be encountered by people attempting to work on the surface of the moon due to

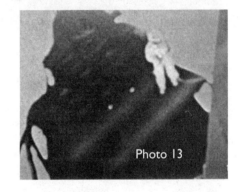

Photo 13

the variety of strange atmospheric and gravitational characteristics that would be encountered. This didn't prove to be the case and after landing we felt very comfortable in the lunar gravity. It was, in fact, in our view preferable both to weightlessness and the earth gravity. All the systems on the LM were operating magnificently. We had very few problems We decided to go ahead with the surface work immediately. We predicted that we might be ready to leave the LM by 8 o'clock, but those of you who followed on the ground recognize we missed our estimate a good deal. This was due to a number of factors, one we had housekeeping to perform and food packages, flight plans, all the items that we had used in the previous descent to be stowed out of the way prior to depressurizing the lunar module. 2 It took longer to depressurize the lunar module than we had anticipated and 3; it also took longer to get the cooling units in our backpacks operating than we had expected. In substance it took us approximately an hour longer to get ready than we had predicted. When we actually descended the ladder it was found to be very much like the lunar gravity simulations we had performed here on earth. No difficulty was encountered in descending the ladder. The last step was about 3 and a half feet from the surface, and we were somewhat concerned that we might have difficulty in reentering the LM at the end of our activity period. So we practiced that before doing the exercise of bringing the camera down which took the subsequent surface pictures. Here you see the camera being lowered on what might be called the Brooklyn clothesline. (Photo 13.) I was operating quite carefully here because immediately to my right and off the picture was a 6 foot deep crater. And I was somewhat concerned about losing my balance on the steep slope. The other item of interest in the very early stages of EVA, should it have been cut short for some unknown reason, was the job of bringing back a sample

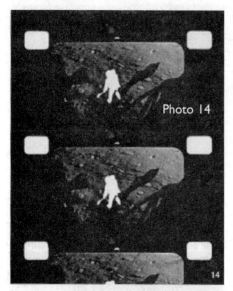

Photo 14

Photo 15

Photo 16

Photo 17

of the lunar rocks. And the photographs show the collection of that initial sample into a small bag. (Photo 14.) And then that bag being deposited in my pocket. This was the first of a number of times when we found (laughter) found 2 men were a great help. I quickly put up the TV camera. (laughter) And then more leisurely Buzz and I joined together to erect the American flag. (Photo 15.) We found on a number of occasions that we were able to help each other in many ways on the surface. You probably recall the times that I got my foot caught in the television cable, and Buzz was able to help me extract it without falling down.

ALDRIN — We had some difficulty at first getting the pole of the flag to remain in the surface. In penetrating the surface, we found that most objects would go down about 5 maybe 6 inches, and then it would meet with a gradual resistance. At the same time there was not much of a support force on either side, so we had to lean the flag back slightly in order for it to maintain this position. So many people have done so much to give us this opportunity to place this American flag on the surface. To me it was one of the prouder moments of my life to be able to stand there and quickly salute the flag. (Photo 16.)

ARMSTRONG — The rest of the activity seemed to go very rushed. (laughter) There were a lot of things to do, and we had a hard time getting them finished.

ALDRIN — We did find that mobility on the surface was in general a good bit better than perhaps we had anticipated it. There was a slight tendency to be more nearly toward the rear of a neutral stable position. Loss of balance seemed to be quite easy to identify. And as one would lean a slight bit to one side or the other, it was very easy to identify when this loss of balance was approaching. In maneuvering around as you saw, this was one of my tasks fairly early in the EVA. I found that a standard loping technique of one foot in front of the other worked out quite well as we would have expected. One could also jump in more of a kangaroo fashion, 2 feet at a time. This seemed to work, but without quite the same degree of control of your stability as you moved along. We found that we had to anticipate 3 to 4 steps ahead in comparison with the 1 or 2 steps that are ahead when you're walking on the Earth.

ARMSTRONG — We had very little trouble. Much less trouble than expected on the surface. It was a pleasant operation. Temperatures weren't high. They were very comfortable. The little EMU, the combination of spacesuit and back pack that provided or sustained our life on surface, operated magnificently. We had no cause for concern at any time with the operation of that equipment. The primary difficulty that we observed was that there was just far too little time to do the variety of things that we would have like to have done. In earlier pictures you saw rocks and the boulder field out Buzz's window that were 3 and 4 feet in size. Very likely pieces of the lunar bed rock. And it would have been very interesting to go over and get some samples of those. There were other craters that differed widely, that would have been interesting to examine and photograph. We had the problem of the 5 year old boy in a candy store. There are just too many interesting things to do. The surface as we said was fine grained with lots of rock in it. It took footprints very well, and the footprints stayed in place. (Photo 17.) The LM was in good shape, and it exhibited no damage from the landing or the descent. Here is a picture of the ladder with the well-known plaque on the primary strut. (Photo 18.) There was a question as to whether the LM would sink in up to its knees. It didn't, as you can see. The footpads sunk in, perhaps, an inch or two. And the probe in this picture was folded over and sticks up through the sand in the bottom right-hand corner (Photo 19.) showing that we were in deep, traveling slightly sidewise, at touchdown. There were a wide variety of surfaces. Here Buzz is standing in a small crater, (Photo 20.) and gives a very good picture of the rounded rims of the - what we believe are very old features. The LM was in a relatively smooth area between the craters and the boulder field. (Photo 21.) And we had some difficulty in determining just what straight up and down was. Our ability to pick out straight up and down was probably several degrees less accurate than it is here on Earth. And it caused some difficulty in having things like our cameras and scientific experiments, sometimes not maintaining a level attitude we expected.

Photo 18

Photo 19

ALDRIN — The two experiments that you saw in the previous picture were deployed in the Scientific Equipment Bay. We found that getting them down produced no significant problem. And here you see a view of my carrying these two experiments out to the deployment site, (Photo 22.) about 70 feet to the south of the lunar module. You have a very good view of the varying depths of the upper surface layer. You see that along the crater rim, a small crater rim off to my left, along this the upper surface appears to be about 2 to 3 inches. The subsurface has a slope that is rather ill-defined, and one has to be very careful in treading your way around these very small craters. Any long excursions, I feel, would take a good bit of attention of your moving along to avoid walking along or down the slope of some of these smaller craters. This is the Passive Seismic Experiment   (Photo 23.) that was deployed and has been giving us good returns on the interactions of the moon.   We had a little difficulty deploying one of the panels. I had to move around to the far side and release the restraining lever, and then the second panel came out. We had a little bit of difficulty determining, as Neil said, the exact local/horizontal, and I think this is due to the decrease in the cues, that a person has as to which way up, up really is. One has to lean a little bit more off to the side before you get this body cue that you're approaching off balance, and of course the terrain

Photo 20

varied considerably in this area. This second experiment is the laser reflector. (Photo 24.) We've been successful in bouncing laser beams off this, from its hundred arrays of reflectors. The other experiment, the solar wind, (Photo 25.) you can see in the background was deployed quite early in the flight and was rolled up, just one of the last things before I reentered the LM. In this picture, you see me driving the core tube into the surface. (Photo 26.) We collected two different core tube samples. It was quite surprising the resistance that was met in this subsurface medium, and at the same time, you see that it did not support very well, the core

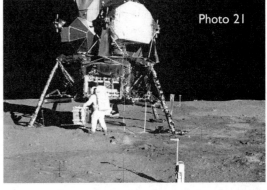

Photo 21

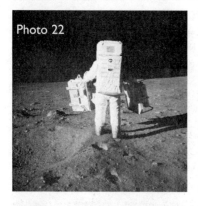

Photo 22

Photo 26

tube as I was driving it into the surface.

ARMSTRONG — This is a close-up picture. (Photo 27.) It's actually a stereo picture of fine particulate material in the moon. This is taken from about an inch or two away from the surface, and shows a shiny coating on some of the clumps there. This appears melted glass and an analysis of the cause for that characteristic is of extreme interest to the scientific community. Second picture taken with that scientific camera shows the nature of the clods of lunar surface material, (Photo 28.) and this picture shows the 80 foot crater, which you observed earlier in the motion picture footage during the final phases of descent. (Photo 29.) We had very much hoped that this crater would be deep enough to show the lunar bedrock. It was about 15 or 20 feet deep, and although there are rocks in the bottom, there is no evidence on the inner walls of the lunar bedrock.

ALDRIN — We deposited several items on the Lunar surface. I'm sure you're aware of these. One was a disc with 73 messages from nations of the world. There was a patch from Apollo 1, and various medals from the cosmonauts. We also elected, as a crew to deposit a symbol which was representative of our patch, that is the US eagle carrying the olive branch to the Lunar surface, and we thought it was appropriate to deposit this replica of the olive branch before we left.

ARMSTRONG — After reentering the LM we could see the effects of our activity on the surface. (Photo 30.) You'll note that the surface looks considerably darker in the area where the majority of the walking took place. However, in the left side of the picture, where it is not as dark, there was also a good bit of walking and so that indicates that the walking probably just increases your ability to notice the effects of the strange lighting that Buzz talked about earlier, where the cross sun lighting is a good bit darker than the down-sun lighting.

Photo 23

Photo 27

Photo 24

Photo 28

Photo 25

Photo 29

ALDRIN — Following the EVA, we had a sleep period, which in a word, didn't go quite as well as we thought it might. We found it was quite difficult to keep warm. When we pulled the window shades over the windows, we found that the environment within the cabin. chilled off considerably and after about 2 or 3 hours, we found that it was rather difficult for us to sleep. You see mounted in the right hand window, the 16 millimeter camera, (Photo 31.) that was mounted for taking the pictures on the surface. Following the sleep period, as we're approaching the lift-off point, we progressed with a gradual power up of the Lunar module, which included another star alignment check, and as Mike came over in Columbia, one revolution before liftoff, we used the radar to track him as he went over. We continued the check out. You see here, one of the data books that's mounted up in front of the instrument panel, (Photo 32.) that was used to record the various messages that were sent up to us, a whole host of numbers, for the particular maneuvers that were coming up, that we would copy down. We would log these on that sort of a data sheet.

Photo 30

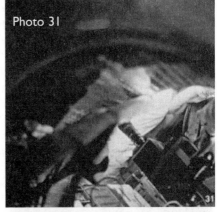

Photo 31

ARMSTRONG — This film shows our final look at Tranquillity Base, (Photo 33.) before our departure, and the ascent was a great pleasure. It was very smooth. We were very pleased to have that engine light up. (laughter) It gave us a excellent view of our take off trajectory, and Tranquillity Base as we left, and at all times through the ascent, we could pick up land marks that assured us that we were on the proper track. There were no difficulties with the ascent and we enjoyed the ride, more than we could say.

ALDRIN — Both guidance systems agreed very closely when we were finally inserted into orbit. I believe they were something on the order of a half a mile, or seven tenths of a mile difference in the apogee, in the resulting orbit. Following an alignment check after insertion into orbit, we proceeded with gathering radar data,

Photo 32

of relative positions between the two vehicles. The solution for the first sequence of rendezvous maneuvers was extremely close and agreed very closely with, with the value that the ground had given us. The surprising feature of this rendezvous, many of us were expecting a fairly large out-of-planeness, due to perhaps some misalignment in azimuth on the surface. We were expecting somewhere up to, maybe 20 or 30 feet per second out-of-plane velocity. We found that we didn't even have to make use of a particular out-of-plane maneuver that had been inserted between two other sequential maneuvers. In comparison with many simulator runs, we found that this was about as perfect a rendezvous as we could have asked for.

Photo 33

ARMSTRONG — You've noted some oscillation in this film during ascent, and that's a real characteristic. The vehicle, due to the changings in our gravity as fuel is used, does a good bit of five degree oscillation throughout the ascent.

COLLINS — This is Eagle, (Photo 34.) and it seems like Columbia, or perhaps half an Eagle would be better since the landing gear and lower part of the descent stage, of course remained on the surface. This was a very happy part of the flight for me. I for the first time, really felt that we were going to carry this thing off at this stage of the game, and it looks like - (laughter) although we were far from home, we were a lot closer to it than the pure distance might indicate. Neil's making the initial, maneuvers

Photo 34

Photo 35

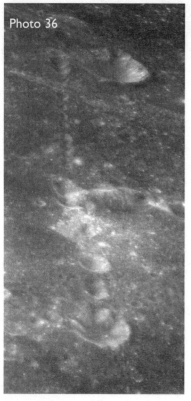

Photo 36

here to get turned around, and then again I do the final docking. This is somewhat swifter than real time. The probe, the dark bundle, on the top of the LM and the docking target below it and to the left in the lighter portion of the LM. The As Buzz said, the rendezvous was absolutely beautiful. They came up from below absolutely as if they were riding on a rail. There was absolutely no disturbance or any off nominal events during the last part of the rendezvous. Upper right you can see the RCS QUADS. And down below the various antenna and other perturbances. This gives you some idea of the rough terrain available on the Moon, (Photo 35.) of course, the Maria on the front side are smoother than this, but in general the back side of the Moon is quite rough and I have, in general, just a series of slides which in the interest of time, I'm not going to dwell on but I just like to point out that we did take a number of pictures I believe from Columbia. We took probably a 1000 stills and some of them show very interesting surface features. Various types of unusual craters and some of them pose many riddles which we hope the geologists will in time be able to answer for us. That line of craters, (Photo 36.) for example, is difficult to explain are at least without an argument it is. Here is a nearer crater with the white material having come from it. (Photo 37.) And this is a picture of the solar corona. (Photo 38.) Neil, would you like to close with that.

ARMSTRONG — During our flight to the moon, we flew through the moon's shadow in fact the moon was eclipsing the sun. We took the opportunity to try to take some photographs of it but our film was just not sufficiently fast to capture the event, however, this does show the brightest part of the solar corona. It extends several moon diameters in addition on each side, they're roughly parallel to that light, but the striking thing to us, as observers, was not the solar corona, but the Moon itself. (Photo 39.) Of course it was dark, un-illuminated by the sun but it was illuminated by the Earth and at this relatively close range it had a decidedly three dimensional effect and was undoubtedly one of the most impressive sites of the flight. As we left the moon, after successful TEI, this is the view that we observed, I think that, at least from where I sit here on the stage, the colors that you see there are quite close to being actually representative of the Moon as seen from that distance. We were sorry to see the Moon go, but we were certainly glad to see that Earth return. (Photo 40.) We took a large number of photographs here on the way out and back and had our wristwatches set on Houston time. An interesting use can be made of that. If you were looking at this picture and you looked at your watch and your watch said 7:00 in the evening, then you'd know that Houston is about 7:00 in the evening and it's about an hour away from sunset so it would be about one twenty-fourth of an earth's circumference away from the shadow, which is just about 15 degrees there, so at anytime by looking at our wristwatch and looking down at the earth, we knew what was underneath the clouds and it aided us in some ways in picking out what we should be seeing. We could see a large number of details on the earth's surface, certainly all the continents and islands and details, many of which you followed perhaps in our discussions over the radio communications but it was interesting to us to find out how well we could observe weather patterns on, not only on the world wide scale that you see here, but in specific localities. This particular shot shows the coast of North America, the equatorial cloud layers, what we think is probably inter tropical conversion zone and cirrus clouds over the Antarctic.

Photo 37

SCHEER — We're ready now for questions and answers and wait for the microphone and we'll go right down the line and we'll catch everyone if you just be patient.

QUERY — How much time did you have left in your life support back packs at the time you got back on board LM?

Photo 38

SPEAKER I haven't seen the post-flight analysis of the numbers, we had roughly half of our available oxygen supply remaining in the back packs and some what less percentage in the water supplies, which is used for cooling. Of course, particularly on our first experience with the use of that back pack on the lunar surface, we were interested in conserving a good bit of margin, in case we had difficulty with closing the hatch or repressuring the LM, or had any difficulties with getting the systems operating again in a normal fashion inside the cockpit.

QUERY — Colonel Aldrin and Mr. Armstrong, when President Nixon made his phone call to you on the Moon, it looked like - the two of you suddenly stopped doing everything and stood there and listened and talked to him. It looked there for a moment like you might be a little bit aware of what was going on. Was there ever a moment on the Moon where either one of you were just a little bit spelled bound by what was going on?

Photo 39

ARMSTRONG — About 2 and a half hours.

QUERY — I'd like to ask Neil Armstrong, when he began to think of what he would say when he put his foot down on the lunar surface and how long he pondered this. The statement about a small step for man, gigantic leap for mankind.

Photo 40

ARMSTRONG — Yes, I did think about it. It was not extemporaneous neither was it planned. It evolved during the conduct of the flight and I decided what the words would be while we were on the lunar surface just prior to leaving the LM.

QUERY — I'd like to ask Neil Armstrong and Buzz Aldrin, and I'm not quite sure how to ask this question, when you first stepped on the Moon, did it strike you as you were stepping - that you were stepping on a piece of the Earth, or sort of what your inner feelings were, whether you felt you were standing on a desert or this was really another world, or how you felt at that point.

ALDRIN — Well, there was no question in our minds where we were, we'd been orbiting around the Moon for some time   At the same time we had experienced one-sixth G before. We've been exposed to some degree to the lighting that we saw.   However, this was, in my case, an extremely foreign situation   with the stark nature of the light and dark condition, and   of course we first step foot on the moon in the dark shadow of the area.

ARMSTRONG — It's a stark and strangely different place, but it looked friendly to me and it proved to be friendly.

QUERY — Some people have criticized the space program as a "Misplaced item on a list of national priorities". I'd like to ask any of the astronauts how do you view space exploration as a relative priority compared with the present needs of the domestic society and the world community at large.

ARMSTRONG — Well, of course we all recognize that the world is continually faced with a large number of varying kinds of problems, and that it's our view that all those problems have to be faced simultaneously, it's not possible to neglect any of those areas, and we certainly don't feel that it's our place to neglect space exploration.

QUERY — All right, there was a lot of discussion during the flight, during the power descent portion of the flight about the program alarms and so forth. I wondered if you all could describe your thoughts on the subject, how it went and what advice YOU might have to offer the crews of Apollo 12 and subsequent flights for this portion of the mission?

ALDRIN — Well, I think we pretty well understand what caused these alarms. It was the fact that the computer was in the process of solving the landing problem and at the same time we had the rendezvous radar in a powered up condition and this tended to add an additional burden to the computer operation. Now I don't think either the ground people or ourselves really anticipated that this would happen. It was not a serious program alarm, it just told us that for a brief instant the computer was reaching a point of being over - programmed or having too many jobs for it to do. Now a computer continually goes through a wait list of one item after another. This list was beginning to fill up and the program alarm came up. Unfortunately it came up when we did not want to be trying to solve these particular problems, but we wanted to be able to look out the window to identify the features as they came up so that we would be able to pin point just where in the landing ellipse stage the computer was taking us.

ARMSTRONG — Suppose we were carrying on a rapid fire conversation with the computer at that point, but we really have to give the credit to the control center in this case. They were the people who really came through and helped us and said "continue". Which is what we wanted to hear.

QUERY — Gentlemen you're about to take some tours. I wonder what your feelings are. Is that perhaps the most difficult part of the mission or are you looking forward to it?

ARMSTRONG — It's certainly the part that we're least prepared to handle.

QUERY — What do you consider the most important piece of advice and recommendation that you will give the Apollo 12 crew before they take off for -she moon in November, gentlemen?

ARMSTRONG — I didn't hear the first part. Recommendations for 12 in which?

QUERY — Which would be the important piece of advice or recommendation for the Apollo 12 crew before they take off for the moon in November, gentlemen?

ARMSTRONG — I think that we can say that overall we wouldn't change the plan that we used or the plan that they intend to use. You know that there are a large number of individual details which we think could stand improvement and we have had the opportunity in the past couple of weeks to go over those details with the crew members and various people from around the program. In general I say that we wouldn't recommend any major changes to the plan.

QUERY — Will you recommend any changes in procedures for the moon walking and exploration procedure and did you find that your suits were mobile enough in view of the changes or would you recommend further mobility features for them for operation on the moon?

ALDRIN — Well, one gets used to the type mobility that your suit affords you and of course we would like to always have more and more dexterity with arms moving and fingers moving. These things are under study. Of course the Apollo 12 mission will have 2 different periods of EVA. One early in the mission and then sleep period and then another EVA following that. We in general looked at their plans and we talked to them about the durations, we talked to them about a brief period at the beginning of their EVA for their familiarization with the EVA the 160 environment. I don't think we have any particular recommendations for how they should change their mission. It is a continuing evolvement of EVA capability and scientific exploration that they're undertaking on that flight.

QUERY — I would like to ask Colonel Aldrin if he would elaborate a little bit on his comment earlier about having to anticipate where you were going to walk 3 or 4 steps in advance as compare to just 1 or 2 on earth. Did you mean that in respect to avoiding craters or deep pits or what?

ALDRIN — Well, I mean it with respect to the inertia that the body has in moving at this rate of 5 to 6 miles an hours that we found to be fairly convenient. Due to the reduced force of gravity your foot does not come down so often, so you have to anticipate ahead and control your body movement and since your foot is not on the surface for a long period of time in each step you're not able to bring to bear large changes in your force application which would enable you to slow down. So in general we found we had to anticipate 3 or 4 steps ahead instead of maybe the 1 or 2 that you do on the surface of the earth...

QUERY — You are now national heroes and you've had a couple of weeks in isolation in the LRL to think about that. What are your initial feelings about being heroes? How do you believe it will change your lives and do you think that maybe you'll get another chance to go to the moon or are you going to be too busy being heroes?

ARMSTRONG — Probably to get an answer to that question might have to spend as long preparing as we had to prepare for Apollo 11. In the Lunar Receiving Laboratory we did have very little time for meditation as it turned out we were quite busy throughout the time period with the same sorts of things that the crews of past flight have done after their flights. The debriefing schedules and writing the pilot reports and getting all the facts down for the use of all the people who will include that in the future flights.

QUERY — I'm struck from the movies and the still pictures by the difference in the very hostile appearance of the moon when you're orbiting over it or some distance from it and the warmer colors and the relatively apparently more friendly appearance of it when you're on the surface. I'd like to ask Colonel Collins if he gets that same impression from the pictures and the 2 of you who were on the moon, what impression do you have along those lines?

COLLINS — The moon changes character as the angle of sunlight striking its surface changes. At very low sun angles close to the terminator at dawn or at dusk. It has the harsh, forbidding characteristics which you see in a lot of the photographs. On the other hand when the sun is more closely overhead the mid-day situation the moon takes on more of a brown color - it becomes almost a rosey looking place - a fairly friendly place so that from dawn through mid-day through dusk you run the whole gamut. It starts off very forbidding, becomes friendly and then becomes forbidding again as the sun disappears.

QUERY — Neil, were you and Buzz - did you get the feeling that you were getting a little low on fuel during landing? Were you concerned at that point about being low on fuel. and the second part of it I suppose for Buzz is, out of your experience how tough do you think that pin point precision landing will be on the lunar surface on future flights?

ARMSTRONG — Yes, we were concerned about running low on fuel. The range extension we did was to avoid the boulder field and craters. We used a significant percentage of our fuel margins and we were quite close to our legal limit.

QUERY — What changes will be based on your experience?

ALDRIN — Well, I think it requires some very pin point determination of the orbit that the vehicle is in before it begins power descent and this requires extreme care in making sure of ground tracking because the entire descent is based upon the knowledge that the ground has and puts into the onboard computer of exactly where the spacecraft is and this starts several revolutions before and then is carried ahead as the computer keeps track of the craft's position. So during sequences like undocking we have to be extremely careful that we do not disturb this knowledge of exactly where it is, because this then relates in the computer to bringing the LM down in a different spot than everyone thought we were coming. This is what defines the error ellipse, where we might possibly land having targeted for the center. Now the ability to be able to control where you are requires that you be able to identify features and, of course, in our particular landing site this was selected to be as void of significant features as possible to give us a smoother surface. In any area like this there are always certain identifying features that you can pick out — certain patterns of craters— to the extent that this can be used. If the crew sees that they are not going exactly toward the pre-planned point, they can begin to tell the computer to move to a slightly different landing location. Now, this can occur up in the region of 5 to 6 thousand feet. Then as Neil took over control of our spacecraft to extend the range, to get beyond this large crater - West crater - this again maybe be required if identification is made in the vicinity of 3, 4 or 5 hundred feet to be able to maneuver that last few seconds in the vicinity of a thousand or 2 thousand feet to make a pinpoint landing. So much depends on the early trajectory, the ability to then redesignate, and the final manual control.

QUERY — For Mr. Armstrong and more on the landing, did at any time you consider an abort while you were getting the alarms and so forth?

ARMSTRONG — Well, I think - in simulations we have a large number of failures and we are usually   spring loaded to the abort position and in this case in the real flight we are spring loaded in the land position. We were certainly going to continue with the descent as long as we could safely do so and as soon as program computer alarms manifest themselves, you realize that you have a possible abort situation to contend with, but our procedure throughout a preparation phase was to always try to keep going as long as we could so that we could bypass these types of problems.

ALDRIN — The computer was continuing to issue guidance throughout this time period and it was continuing to fly the vehicle down in the same way that it was programmed to do. The only thing that was missing during this time period is that we did not have some of the displays on the computer keyboard and we had to make several entries at this time in order to clear up that area.

QUERY — Would the crew consider a moon mission of a similar nature again or would you prefer to have some other kind of mission and secondly, I think this question was asked, but I did not get the complete answer. How do you propose to restore some normalcy to your private lives in the years ahead ?

ALDRIN — I wish I knew the answer to the latter part of your question.

ARMSTRONG — It kind of depends on you. But I think that the landings that are presently considered for the next number of flights are appropriate to the conclusions that we reached as a result of our descent. I would   certainly hope that we are able to investigate the variety of types of landing sites that they hope to accomplish.

QUERY — I have 2 brief questions that I would like to ask, if I may. When you were carrying out that incredible moon walk, did you find that the surface was equally firm everywhere or were there harder and softer spots that you could detect. And, secondly, when you looked up at the sky, could you actually see the stars in the solar corona in spite of the glare?

ALDRIN — The first part of your question, the surface did vary in its thickness of penetration somewhere in flat regions. The footprint would penetrate a half an inch or sometimes only a quarter of an inch and gave a very firm response. In other regions near the edges of these craters we would find that the foot would sink down maybe 2, 3, possibly 4 inches and in the slope, of course, the various edges of the footprint might go up to 6 or 7 inches. In compacting this material it would tend to produce a slight sideways motion as it was compacted on the material underneath it. So we feel that YOU cannot always tell by looking at the terrain what the exact resistance will be as your foot sinks into a point of firm contact. So one must be quite cautious in moving around in this rough terrain.

ARMSTRONG — We were never able to see stars from the lunar surface or on the daylight side of the moon by eye, without looking through the optics. I don't recall during the period of time that we were photographing the solar corona what stars we could see.

ALDRIN — I don't remember seeing any.

QUERY — Neil, you were a little bit concerned you said about stubbing your toe at the point of landing because the surface was obscured by dust, do you see any way around that problem for future landings on the moon?

ARMSTRONG — I think that the simulations that we have at the present time to enable a pilot to understand the problems of a lunar landing, that is, the simulator and the various lunar landing training facilities and trainers that we have will do that job sufficiently well. Above that, I think it is just a matter of pilot experience.

QUERY — This is For Neil Armstrong. You said earlier in your presentation that Maskelyne W occurred about 3 seconds late giving you the clue that you might land somewhat long. Now this was before you got the high gate so that it had nothing to do with maneuvering to find a suitable place to land. I am wondering what would have caused this 3 seconds late. Did it have something to do with the time that you began the powered descent or what?

ARMSTRONG — The time that we started powered descent was the planned time but the question is where are you over the surface of the moon at the time of ignition and where that point is, is largely determined by a long chain of prior events, tracking that has taken place several revolutions earlier, the flight maneuvers that have been done in checking out the rate control systems, the undocking and the ability to station-keep accurately without ever flying very far away from where the computer thinks you ought to be at that time. And, of course, the little bit of dispersions in a maneuver such as the deal I burned on the back side of the moon that were not quite properly measured by the guidance system. Each of those things will accumulate into an effect that is an error - a position error - at ignition and there is no way of compensating until you get to final phase for that error.

QUERY — Based on your own experience in space, do you or any of you feel that there will ever be an opportunity for a woman to become an astronaut in our space program?

ARMSTRONG — Gosh, I hope so.

QUERY — I would like to refer back to something that Neil Armstrong said a while back, that there was so much other he would liked to have done. As it was, you ended up a considerable number of minutes behind the schedule. Is that because the schedule was overloaded for the EVA or can we expect all astronauts, when they reach the moon for the first time, to enjoy themselves and spend as much time doing so as you seemed to?

ARMSTRONG — We plead guilty to enjoying ourselves. As Buzz mentioned earlier, we are recommending that we start future EVA's with a 15 or 20 minute period to get these kinds of things out of the way and to get used to the surface and what you see, adapt to the 1/6 g in maneuvering around and probably we just included a little more in the early phase than we were actually able to do.

QUERY — — Two questions. Where did the weird sounds including the sirens and whistles come from during the transearth coast. I believe ground control had asked for explanations saying it had come from the spacecraft. Secondly, I understand that although low angle lighting caused no problem walking around, there was a problem seeing obstacles in time when traveling at high speeds. I understand this might indicate the need for flying machines rather than a rover for long distance lunar surface travel. Can you explain this?

ARMSTRONG — We are guilty again. We sent the whistles and — (Laughter) and bells — with our little tape recorder which we used to record our comments during the flight in addition to playing music in the lonely hours. We thought we'd share that with the people in the Control Center. The (laughter) sun angle was less a problem for the things you mentioned than the lunar curvature and the local roughness. It seemed to me as though it was like swimming in an ocean with 6 foot or 8 foot swells and waves. In that condition, you never can see very far away from where you are. And this was even exaggerated by the fact that the lunar curvature is so much more pronounced.

QUERY — This is for Mr. Armstrong. Had you planned to take over semi-manual control, or it was only your descent toward the West crater that caused you to do that?

ARMSTRONG — The series of control system configurations that were used during the terminal phase were in fact very close to what we would expect to use in the normal case, irrespective of the landing area that you found yourself in. However, we spent more time in the manual phase than we would have planned in order to find a suitable landing area.

QUERY — Many of us and many other people in many places have speculated on the meaning of this first landing on another body of space. Would each of you give us your estimate of what is the meaning of this to all of us?

ARMSTRONG — You want to try it?

ALDRIN — After you.

ALDRIN — Well, I believe, that what this country set out to do was something that was going to be done sooner or later whether we set a specific goal or not. I believe that from the early spaceflights, we demonstrated a potential to carry out this type of a mission. And again it was a question of time until this would be accomplished. I think, the relative ease with which we were able to carry out our mission which, of course, came after a very efficient and logical sequence of flights.... I think that this demonstrated that we were certainly on the right track when we took this commitment to go to the moon. I think that what this means is that many other problems perhaps can be solved in the same way by taking a commitment to solve them in a long term fashion. I think, that we were timely in accepting this mission of going to the moon. It might be timely at this point to think in many other areas of other missions that could be accomplished.

COLLINS — To me there are near and far term aspects to it. On the near term, I think it a technical triumph for this country to have said what it was going to do a number of years ago, and then by golly do it, just like we said we were going to do. Not just, perhaps, purely technical, but also a triumph for the nations overall determination, will, economy attention to detail, and a thousand and one other factors that went into it. That's short term. I think, long term, we find for the first time that man has the flexibility or the option of either walking this planet or some other planet, be it the moon or Mars, or I don't know where. And I'm poorly equipped to evaluate where that may lead us to.

ARMSTRONG — I just see it as beginning. Not just this flight, but in this program which has really been a very short piece of human history, an instant in history, the entire program. It's a beginning of a new age.

QUERY — Neil, how much descent fuel did you have left when you actually shut down?

ARMSTRONG — My own instruments would have indicated less than 30 seconds. Probably something like 15 or 20 seconds, I think, the analyses made here on the ground indicate something more than that. Probably greater than 30 seconds; 40 or 45. That sounds like a short time but it really is quite a lot.

QUERY — This is for Colonel Collins. You used a rather colorful expression when there seemed to be some problem with docking. Could you tell us precisely what was going on at that time? Were you docked and then...

COLLINS — Are you referring to the lunar orbit docking when after the 2 vehicles made contact, a yaw oscillation developed. This oscillation covered, perhaps, 15 degrees in yaw over a period of 1 or 2 seconds and was not normal. It was not anything that any of us expected. It was not a serious problem. It was all over in an additional 6 or 8 seconds. The sequence of events is that the 2 vehicles are held together initially by 3 capture latches and then a gas bottle, when fired, initiates a retract cycle which allows the two to be more rigidly connected by 12 strong latches around the periphery of the tunnel. Now this takes 6 or 8 seconds for this cycle, between initial contact and the retract. And it was during this period of time, that I did have a yaw oscillation, or we did. Neil and I both took manual corrective action to bring the 2 vehicles back in line. And while this was going on the retract cycle was successfully taking place. And the latches fired, and the problem was over.

QUERY — Two questions Col. Aldrin, the pictures taken on the surface, your fold portrait, shows the distinct smudges of lunar soil on your knees. Did you fall down on the surface or kneel? And then for Mr. Armstrong, during the last few minutes there, before the landing when the program alarms were coming on and et cetera, would you have gone ahead and landed had you not had ground support?

ALDRIN — To my recollection, my knees did not touch the surface at any particular time. We did not feel that we should not do this.   We felt that this would be quite a natural thing to do to recover objects from the surface, but at the same time we felt that we did not want to do this unless it was absolutely necessary. We found quite early in the EVA that the inter surface material did tend to adhere considerably to any part of the clothing. It would get on the gloves and would stay there. When you would knock either your foot or your hand against something, you would tend to shed the outer surface of this material, but there remained considerable smudges. I don't know how that got on the knees.

ARMSTRONG — Neither of us fell down. We would have continued the landing, so long as the trajectory seemed safe. And a landing is possible, under these conditions, although with considerably less confidence than you have when you have the information from the ground and the computer in its normal manner being available to you.

QUERY — For Mr. Armstrong and Col. Aldrin. Would you please give us a bit more detail about your feelings, your reactions, your emotions during that last several hundred feet of powered descent? Especially when you discovered that you were headed for a crater full of boulders and had to change your landing spot.

ARMSTRONG — Well, first say that I expected that we would probably have to make some local adjustments to find a suitable landing area. I thought it was highly unlikely that we would be so fortunate as to come down in a very smooth area, and we planned on doing that. As it turned out, of course, we did considerably more maneuvering close to the surface than we had planned to do. And the terminal phase was absolutely chock full of my eyes looking out the window, and Buzz looking at the computer and information inside the cockpit and feeding that to me. That was a full-time job.

ALDRIN — My role during the latter 200 feet is one of relaying as much information that I can that is available inside the cockpit in the form of altitude, altitude rate, and forward or lateral velocity. And it was my role of relaying this information to Neil so that he could devote most of his attention to looking out. What I was able to see in terms of these velocities and the altitudes appeared quite similar to the way that we had carried out the last 200, 100 feet in many of our simulations.

SCHEER — We are going to end here. We would appreciate it very much if you would remain seated so that the crew could leave. They want to be at the luncheon at 1 o'clock, just slightly ahead of you. Thank you very much.

Michael Collins during a simulation inside the
CM. The only pictures of Collins taken during
the flight were video and movie.

The crew receive a heroes welcome in New York City. — Summer 1969